Building a Quadcopter with Arduino

Design, build, and fly your personal quadcopter, using commonly available electronics to capture incredible photographs and record phenomenal videos

Vasilis Tzivaras

[PACKT] open source
community experience distilled
PUBLISHING

BIRMINGHAM - MUMBAI

Building a Quadcopter with Arduino

First published: December 2015

Production reference: 1221215

Published by Packt Publishing Ltd.
Livery Place
35 Livery Street
Birmingham B3 2PB, UK.

ISBN 978-1-78528-184-6

www.packtpub.com

Credits

Author
Vasilis Tzivaras

Reviewer
Joseph Monti

Commissioning Editor
Priya Singh

Acquisition Editors
Usha Iyer
Reshma Raman

Content Development Editor
Arshiya Ayaz Umer

Technical Editor
Taabish Khan

Copy Editors
Rashmi Sawant
Sneha Singh

Project Coordinator
Shipra Chawhan

Proofreader
Safis Editing

Indexer
Priya Sane

Production Coordinator
Shantanu N. Zagade

Cover Work
Shantanu N. Zagade

About the Author

Vasilis Tzivaras is a software developer and hardware engineer who lives in Ioannina, Greece. He is currently an undergraduate student of the Department of computer science and Engineering at Ioannina. Alongside his studies, he is working on many projects relevant to robotics, such as drones, home automation, and smart home systems using Arduino and the Raspberry Pi. Furthermore, he is enthusiastic about clean energy solutions and cultural innovation ideas.

He has worked for the University Hospital of Ioannina as an assistant for various computer issues, and he has been part of the support team of his CSE department for over a year. He has participated in IEEE UOI Student Branch and other big organizations, such as FOSSCOMM, with personal presentations for website designing, programming, Linux systems, and drones.

As a member of IEEE University of Ioannina Student Branch, he has proposed many projects and solutions to automate homes and many other life problems by reducing the time of everyday routines. In addition to this, he has come up with ideas to entertain kids with funny and magical projects using Arduino-like hardware and open source software. Many of the projects can be found at his GitHub account under the name of *BillyTziv*.

Apart from this book, he has also worked on another book *Programming in C,* which is not yet published. In addition to this, he has written for blogs, forums, guides, and small chapters, explaining and sharing his knowledge of computers, networks, and programming.

Firstly, I would like to thank my entire family and my close friends, Joe Polenakis, Marios Balamatsias, and Kostis Karantias for supporting me every single day. Secondly, I could do nothing without Vasilis Megas who stood by me from the first day, assisting me while building my first drone and giving me instructions about safe flights. Furthermore, I would like to thank my advisor, Kostas Vlaxos, for helping me out with my research about drones. Lastly, everyone else who said that I could not do this; they set me a challenge, which I finally accepted.

About the Reviewer

Joseph Monti is a software engineer who has worked in the field of software engineering for more than 12 years. His programming languages of choice are Scala and Python. He is currently the director of software engineering at ZenoRadio, which combines telephony and internet-based radio to connect listeners to broadcasters. His range of expertise includes web services, data processing systems, ad serving technology, and robotics. Whether it is an algorithm, data model, or software architecture, he relishes a challenge.

From an early age, Joseph's passion for technology motivated him to build and race radio controlled cars and program his calculator to do his homework. His interests in both hardware and software lead him to the field of robotics, and more recently, quadcopters, AI, and machine learning. He is currently working on a quadcopter as a prototype platform, controlled by Arduino and the Raspberry Pi. While he enjoys flying quadcopters, this project is intended for the challenge of overcoming the complexities of flights and taking advantage of the freedom it allows. His website and blog can be found at `http://joemonti.org/`.

www.PacktPub.com

Support files, eBooks, discount offers, and more

For support files and downloads related to your book, please visit www.PacktPub.com.

Did you know that Packt offers eBook versions of every book published, with PDF and ePub files available? You can upgrade to the eBook version at www.PacktPub.com and as a print book customer, you are entitled to a discount on the eBook copy. Get in touch with us at service@packtpub.com for more details.

At www.PacktPub.com, you can also read a collection of free technical articles, sign up for a range of free newsletters and receive exclusive discounts and offers on Packt books and eBooks.

PACKTLiB™

https://www2.packtpub.com/books/subscription/packtlib

Do you need instant solutions to your IT questions? PacktLib is Packt's online digital book library. Here, you can search, access, and read Packt's entire library of books.

Why subscribe?

- Fully searchable across every book published by Packt
- Copy and paste, print, and bookmark content
- On demand and accessible via a web browser

Free access for Packt account holders

If you have an account with Packt at www.PacktPub.com, you can use this to access PacktLib today and view 9 entirely free books. Simply use your login credentials for immediate access.

Table of Contents

Preface

Quadcopters have made an unexpected comeback from 1907, where all the research about multicopters paused due to stability issues. Nowadays, this kind of aerial vehicle is the perfect solution for many applications. If you want to capture a photo or record a video, transfer low-weight payload quickly and securely, or join races with other hobbyists, then you will have to buy one or even create one on your own.

As an author, I have tried to keep the simplicity of the quadcopter but also give plenty of instructions on how you can easily create one for your own application. As kids, we always wanted something that could fly and that we could play around with. Through this book, this dream can come true.

What this book covers

Chapter 1, Introduction to Quadcopters, provides information on various categories of drones, and you will be able to understand why a quadcopter is stable and can hover. Furthermore, some possible applications will be mentioned as well.

Chapter 2, Hardware Overview, analyzes every part of the quadcopter and gives some basic instructions about the usage of each component.

Chapter 3, Creating a Frame, analyzes some possible materials and the shapes and purposes of different frames, since each component will be mounted on a frame. Some frames are designed for stability purposes and photography, whereas others are used for quadcopter races.

Chapter 4, Soldering the Electronics, covers how to solder wires and create connections between different parts, as some of the electronics require further modifications as per individual requirements.

Chapter 5, Electronics Installation, contains some tips for better stability even though installing the electronics is a piece of cake. Notice that it's quite easy to damage something and have a problem later.

Chapter 6, Flight Controller Setup, covers how to set up ArduPilot, which is an open source and easy flight controller. A flight controller is the core component of your quadcopter. It contains all the software that is crucial for stability. Setting up a controller may be easy or very hard.

Chapter 7, Flight Instructions, helps you ask some basic questions and tries to make you think like a professional pilot. Your quadcopter is ready but flying is not so easy. There are many factors that you should think of before going out there and powering up your quadcopter; where you will fly, what will happen if something goes wrong, how much time you have, and so on. All these queries should be answered before flying.

Chapter 8, Cameras and Extra Functions, covers vibration issues and mechanisms that you can use or build on your own. Apart from all the basic components, there are more advanced ones that you can easily add to your quadcopter, depending on the purpose of the flight. FPV is a great way to have eyes in the sky and LEDs are necessary for night flights.

Chapter 9, Repair and Maintenance, covers how to maintain your vehicle, how to protect it from wear, and why this is crucial for safe flights. Even professionals face a crash. This is something that you will unfortunately experience, but the goal here is to estimate the damage and repair your quadcopter as fast as you can.

What you need for this book

This book is intended for hobbyists who are looking forward to fulfilling their dream of flying a quadcopter made by them. Some basic experience in electronic circuits and aerial vehicles will make it easier for you; however, there is nothing too complicated or difficult to do if you follow all the instructions sequentially.

Who this book is for

This book is for everyone who has had the dream of flying their own vehicle or building an autonomous drone. Obviously, hobbyists will understand the instructions better than other people. It actually depends on your knowledge, but if you follow every instruction, you should be able to make it fly easily.

Conventions

In this book, you will find a number of text styles that distinguish between different kinds of information. Here are some examples of these styles and an explanation of their meaning.

Code words in text, database table names, folder names, filenames, file extensions, pathnames, dummy URLs, user input, and Twitter handles are shown as follows: "ArduPilot is connected to COM5."

New terms and **important words** are shown in bold. Words that you see on the screen, for example, in menus or dialog boxes, appear in the text like this: "As soon as you click on **CONNECT**, there are four things that will happen."

> Warnings or important notes appear in a box like this.

> Tips and tricks appear like this.

Reader feedback

Feedback from our readers is always welcome. Let us know what you think about this book—what you liked or disliked. Reader feedback is important for us as it helps us develop titles that you will really get the most out of.

To send us general feedback, simply e-mail `feedback@packtpub.com`, and mention the book's title in the subject of your message.

If there is a topic that you have expertise in and you are interested in either writing or contributing to a book, see our author guide at `www.packtpub.com/authors`.

Customer support

Now that you are the proud owner of a Packt book, we have a number of things to help you to get the most from your purchase.

Downloading the color images of this book

We also provide you with a PDF file that has color images of the screenshots/diagrams used in this book. The color images will help you better understand the changes in the output. You can download this file from `https://www.packtpub.com/sites/default/files/downloads/BuildingAQuadcopterWithArduino_ColorImages.pdf`.

Errata

Although we have taken every care to ensure the accuracy of our content, mistakes do happen. If you find a mistake in one of our books—maybe a mistake in the text or the code—we would be grateful if you could report this to us. By doing so, you can save other readers from frustration and help us improve subsequent versions of this book. If you find any errata, please report them by visiting `http://www.packtpub.com/submit-errata`, selecting your book, clicking on the **Errata Submission Form** link, and entering the details of your errata. Once your errata are verified, your submission will be accepted and the errata will be uploaded to our website or added to any list of existing errata under the Errata section of that title.

To view the previously submitted errata, go to `https://www.packtpub.com/books/content/support` and enter the name of the book in the search field. The required information will appear under the **Errata** section.

Piracy

Piracy of copyrighted material on the Internet is an ongoing problem across all media. At Packt, we take the protection of our copyright and licenses very seriously. If you come across any illegal copies of our works in any form on the Internet, please provide us with the location address or website name immediately so that we can pursue a remedy.

Please contact us at `copyright@packtpub.com` with a link to the suspected pirated material.

We appreciate your help in protecting our authors and our ability to bring you valuable content.

Questions

If you have a problem with any aspect of this book, you can contact us at `questions@packtpub.com`, and we will do our best to address the problem.

1 Introduction to Quadcopters

A quadcopter is an aircraft lifted and propelled by four horizontal rotors; each rotor consists of two or three rotor blades. Quadcopters are classified as rotorcrafts or a rotary-wing aircraft to distinguish them from fixed-wing aircraft because the quadcopter derives its source of lift from the rotor blades rotating around a mast. They belong to a larger category of aircrafts defined as multicopters.

Multicopters have various names based on the number of propellers mounted on the craft. A four-propeller multicopter is called a quad-copter, a six propeller multicopter is called a hexa-copter, and nowadays we even have octo-copters.

Over the last few years, remote control airborne vehicles have been incredibly popular for hobbyists, academic research groups, and for individuals who are planning for a start-up. The names of these vehicles vary from quadcopters to quadrotors or even drones. However, they refer to the same vehicle and the most appropriate nomenclature for these crafts is quadrotors. For the sake of simplicity, we will call them quadcopters.

This book will cover the basics of a quadcopter and will guide you on how to build a simple and fully functional quadcopter that is equipped with a camera. Hopefully, by the end of the book you will be able to fly your own quadcopter and capture pictures of your favorite places. All the chapters are necessary for a successful flight and some chapters have instructions on how you can fly your quadcopter, where you should fly, and tips about safety. You should follow all the instructions mentioned and do not skip any of them, even if you feel that a chapter is simple and you have nothing to gain from it.

In this chapter, we will cover the following topics:

- The basic knowledge to get familiar with quadcopters
- Applications of copters
- Comparison with other devices

Basic components

A quadcopter is an aerial vehicle that consists of mechanical and electrical components including the frame, motors, and other electronic parts. It usually consists of the following parts:

- Motors
- Propellers
- **Electronic speed controllers** (ESC)
- Battery and a **power distribution board** (PDB)
- **Flight controller** (FC)
- Frame

More details about the electronic parts, connections, and the frame design will be given in *Chapter 2, Hardware Overview*, and *Chapter 3, Creating a Frame*, respectively.

Basic categories

Almost all quadcopters are **unmanned aerial vehicles** (**UAV**) or drones. This means that the craft is controlled by a pilot on the ground or in another vehicle. **Remotely piloted aircraft** (**RPA**) is another definition that can be used. However, we can separate UAVs into three categories according to the level of **artificial intelligence** (**AI**) their craft flight controller has.

Fully remote-controlled vehicles (fully RCV)

This category of craft is fully controlled by a pilot using a ground station. Crafts of this type are preferred by hobbyists.

Hybrid remote-controlled vehicles (hybrid RCV)

It is possible to increase the level of intelligence of a UAV so that it will neither be a fully autonomous vehicle, nor a fully remote controlled one. The pilot uses a remote control device to communicate between the quadcopter and the ground station but now the pilot does not fly the craft as if it's an RCV, the only thing that is required now is to set the points that it should pass (navigation points).

Autonomous flying vehicles

This category of craft has no pilot. All we need to do is connect the battery to the electronic system and let it fly. Quadcopters have numerous sensors, such as GPS, accelerometers, cameras, and many more. Every second, the controller gathers data from all the sensors and after calculations it autonomously decides how and where it should go according to the mission plan. These are extremely difficult to build, which is the reason why we prefer hybrid RCVs if we need a craft with autonomous functions.

Different frame shapes

Usually, any inexperienced hobbyist who is willing to build a quadcopter chooses the **X** shape, with four motors mounted at the ends. There are plenty of shapes and materials that can be used for this construction. Each has its own advantages and drawbacks, as it will be explained later on in the subsequent chapters. The X shape is the most common and a first attempt for almost everyone. However, a quadcopter can also be created in a **Plus** or **H** shape, as shown in the following figure:

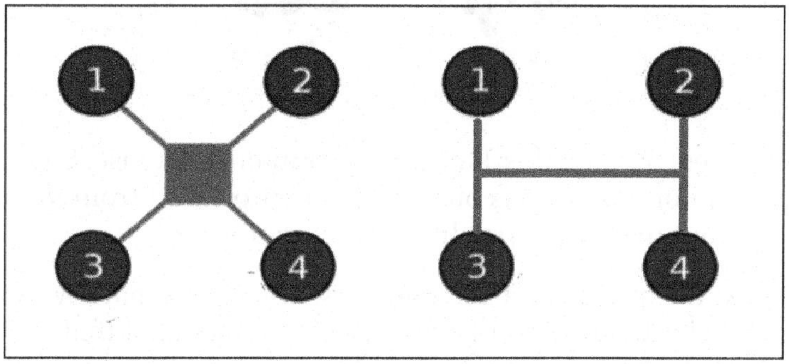

Adding more than four motors will result in a different build that might also work. Vehicles like these are defined as hexa-copters or octo-copters depending on the number of motors used. It should be clear that we can only add two motors at a time; adding only one will not work due to the laws of physics, as will be explained later.

Flight dynamics

A quadcopter consists of four motors. Each motor produces thrust and torque. The quadcopter can hover and stay steady over the ground when all four motors have the same torque and the total thrust of the craft is equal to or greater than $m*g$, where m is the mass of the craft and g is the gravity constant. Motors one and four spin clockwise, as shown in the following figure:

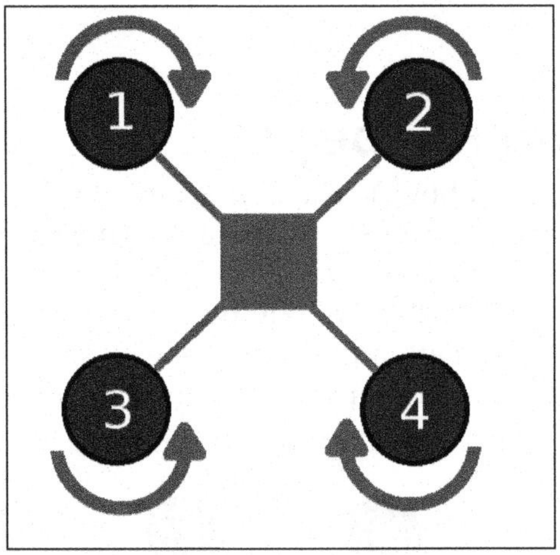

Motors two and three, on the other hand, spin counter clockwise. Since every motor produces torque, motor one cancels out the torque produced from motor four and motor two cancels out the torque produced from motor three.

Let's have a look at the quadcopter's maneuvers. Assuming that *CM* is the center of the craft mass; in order to implement a forward movement, all that is needed is to lower the angular velocity of motor one and two. As a result, there will be less thrust produced in the front than the rear resulting in the quadcopter moving forward. Similarly, we can lower the angular velocity of motors one and three or three and four and have either left or right movement. Lastly, there is one more maneuver that a quadcopter can do. In order to have yaw rotation, we need to lower the speed of any of the two opposite motors. Actually, doing this will increase or decrease the total left or right torque. This creates an imbalance in the torque between the motors; thus, causing the quadcopter to have yaw rotation.

Applications

The world is in need of something fast, light, and reliable. This technology is already developed but it is not yet reliable. Therefore, it is too early to use quadcopters in our everyday life. However, they can be used in several other ways.

Research platform

Quadcopters are quite interesting. They are used as a research platform for individual researchers and university research groups, to test and evaluate new ideas in a number of different fields, including flight control theory, navigation, real time systems, robotics, and so on. Researchers usually build, modify, or observe their maneuvers to develop more stable and promising future crafts. Being a researcher for a system like this requires a very good knowledge about the dynamics and mechanics of aerial vehicles. Furthermore, test flights cannot be done in real time. Simulation software is an important tool, so that researchers can test their algorithms and theories. Anything that passes the simulation stage will be tested in the simulation indoor lab where safety is a number one priority.

Swarms of quadcopters can hover in formation autonomously, perform flying routines, such as flips, dart through hula hoops and organize themselves to fly through windows as a group. They are relatively cheap, available in a variety of sizes, and their simple mechanical design means that they can be built and maintained by amateurs.

As they are so maneuverable, quadcopters can be useful in all kinds of situations and environments. Quadcopters capable of autonomous flight can help remove the need for people to put themselves in a number of dangerous positions. This is a prime reason why the research interest has been increasing over the years. For example, a mountain climber can deploy a quadcopter autonomously and check whether he should be more careful or not while climbing. Another idea is to deploy a quadcopter in a house to check whether there are human survivors or not in case of a disaster. Autonomous operations are extremely useful in situations such as earthquakes, fires, and so on.

There are several engineering research laboratories that are currently developing more advanced control techniques and applications for quadcopters. These include mainly MIT's Aerospace Controls Lab, ETH's Flying Machine Arena, and University of Pennsylvania's **General Robotics, Automation, Sensing and Perception** (**GRASP**) Lab.

Commercial use

By the end of 2014, quadcopters were released in the market for the general public. As a result, crafts like quadcopters are now used to capture images and videos of areas that are rough or dangerous for humans. There are many off-the-shelf quadcopters that can follow a transmitter bracelet, which can be worn on the arm. There are quadcopters that can follow subjects without bracelets, using image recognition systems such as OpenCV. The most common commercial quadcopter is one of the copters shown in the following screenshot, and is named **phantom** from DJI and has already made hundreds of thousands of dollars in sales all over the world:

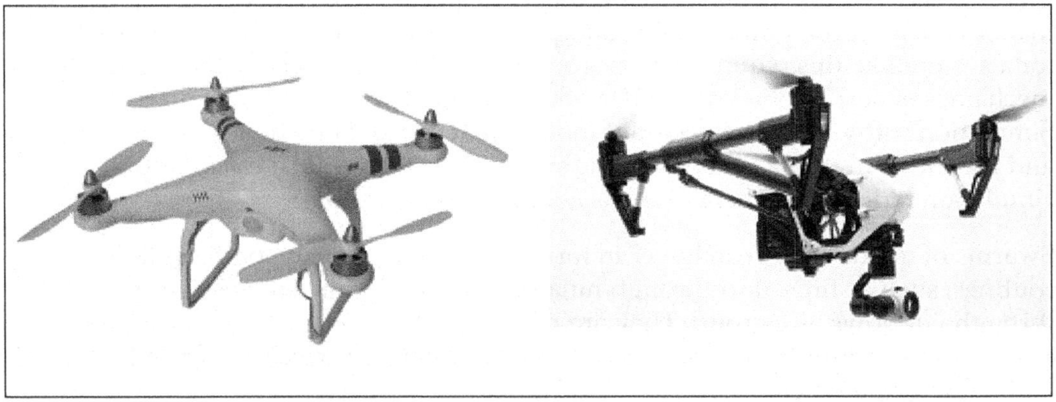

Military and law enforcement

Since UAVs are great for surveillance, quadcopters are now used by military and law enforcement agencies to search, spy, or even rescue teams in missions. There are projects that are being run by militaries for autonomous indoor 3D scanning.

Future potential

Quadcopters can also save lives! After an earthquake or the collapse of a building, which might cause significant destruction, the response time to search and locate trapped survivors is crucial. Mini and fast quadcopters equipped with cameras can fly among the debris in a safe, cheap, and effective way and locate survivors quickly.

Another scenario worth looking at is people drowning in a sea or lake. Every year, the percentage of accidents is almost the same but now quadcopters can change that. There is already an idea of a quadcopter equipped with three buoys and able to release them over three possible targets. The quadcopters will be fully remote controlled by a pilot that overviews the whole mission. So, in each mission a quadcopter will be able to increase the probability of a rescue.

Lastly, quadcopters are excellent aerial vehicles for transportation of low weight cargo. They can be deployed easily in rough areas or big cities. There is a huge need for a speedy vital organ transportation system in various areas, particularly in Africa where the cost of transportation by road is high and extremely difficult.

Comparison with other aerial vehicles

There are various other aerial vehicles that can be used. Humans have successfully made helicopters and airplanes for transportation, surveillance, and even war. Can Quadcopters be used for such causes? Are there any other benefits? Obviously, each vehicle has its own advantages and disadvantages.

Difference between quadcopters and airplanes

Quadcopters and almost every other multicopter craft uses its motors and propellers to produce the necessary thrust needed for the lift, whereas airplanes use motors to produce speed. The lift is achieved by the aerodynamic model of the plane and its wings. So the asset of a quadcopter is the stability that it offers; thus providing a great opportunity to capture photos and videos. On the other hand, airplanes are much faster because of their aerodynamic design and as a result they need less power than helicopters and quadcopters, where all the lift is produced by the motors.

Difference between quadcopters and helicopters

The main difference between a quadcopter and a helicopter is the number of blades used for the lift. A helicopter's mechanism is extremely difficult to build and work properly, whereas quadcopters are pretty simple. A helicopter produces all the lift needed, which is equal to its weight, with only one blade. As a result the total mechanism is quite complex. On the other hand, the stability in a quadcopter comes from four motors with their blades and it's quite simple to understand and create. Thus, the mechanism is quite simple. When a blade is spinning, there are two things that happen. First of all, there is thrust, which lifts the craft. Secondly, every motor creates torque, which is why a craft with only one blade would spin around itself. In a helicopter, a rear blade is used to prevent the craft from spinning around itself. On the other hand, in a quadcopter, all the lift is produced by the sum of each motor's lift, which must be equal to the quadcopter's weight. But, instead of a rear motor to prevent the craft from spinning around, each motor eliminates the torque that the opposite motor produces. As a result, the quadcopter hovers steady without spinning around itself.

Summary

It can be seen from the preceding analysis that there are numerous applications for a quadcopter. In the upcoming chapters, we will build a quadcopter with four motors mounted on an X-shape frame. Even if almost all the references will be to specific parts, as shown in every figure, any other component with similar characteristics should be fine. In the next chapter, we will go through all the crucial parts of the quadcopter and describe each one's usage in the overall build. By the end of the chapter, it should be clear how the quadcopter works as far as the electronic parts are concerned.

2

Hardware Overview

In this chapter, we will go through almost every crucial component needed for the build. All the parts will be explained and information about the usage of each one will be given. Furthermore, you will be able to find some simple instructions and a list of specifications that you should be aware of when buying a component. Thus, you will know how to choose your component based on your requirements. Hopefully, by the end of the chapter you will be able to estimate the cost of your build and have a total overview. This is extremely important when building a craft because you always need to be aware of the final purpose. You always need to know and have an abstract picture of the parts you need and how they will be connected.

In this chapter, we will cover the following topics:

- The motors and propellers
- The **electronic speed controller** (ESC)
- The flight controller
- The battery and charger
- The power distribution system

Radio transmitter and receiver (TX/RX)

It is important to have an overview of the connections between the battery, motors, and the ESCs. The following figure will make this clear:

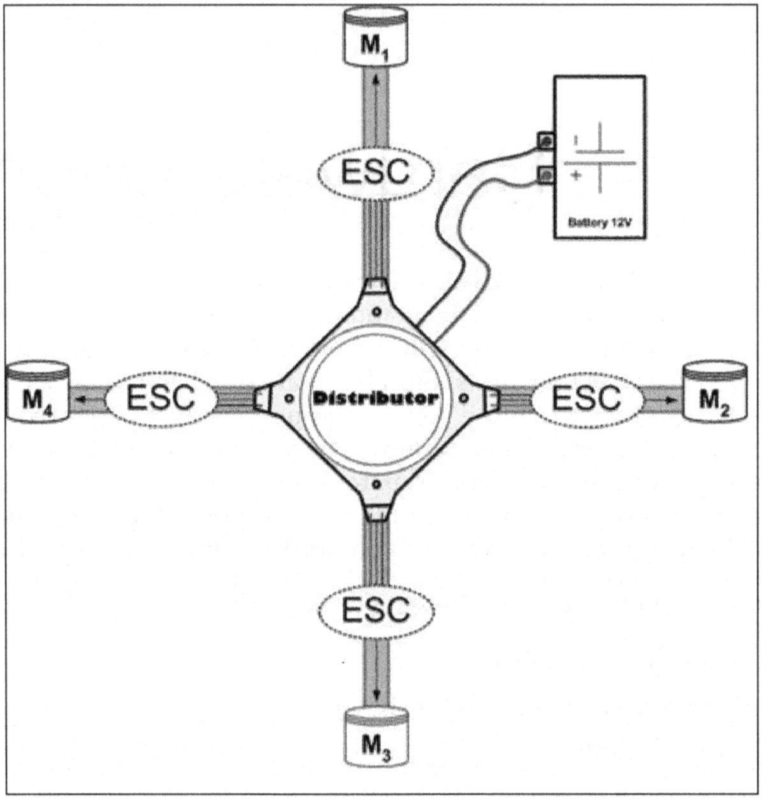

Motors and propellers

A propeller is mounted on the shaft of the corresponding motor. The total number of motors that are needed depends on the build. For example, quadcopters use four motors, hexacopters use six, octocopters use eight, and so on.

Nowadays, **brushless** out runner motors are the most common type of motors used; they have minimal friction. A cylindrical shell of magnets rotates on precision bearings around a core of tightly and neatly coiled wire. The propeller is mounted at the top and is rotating at the same angular velocity as the motor. As a result, thrust is produced and the quadcopter lifts from the ground. Every motor requires reasonable maintenance; dirt should be kept out of the motor so that it can spin easily.

http://www.emaxmodel.com

Common motor specifications

For every motor, there is a specification manual either provided from the manufacturer or the seller. The following is a list of specifications that almost every motor has:

- KV (rpm/v)
- Weight (g)
- Max current (A)
- Max voltage (V)
- Length (mm)
- Shaft (mm)
- Diameter (mm)

An example of the specifications that can be found when choosing a motor is shown in the following table. These specifications were on the same webpage where the motor was bought from. Furthermore, motors typically specify a recommended battery cell count, for example 3S, 4S, or even 5S. It is important to buy recommended parts if you are not an expert:

KV	935
Weight	55G
Diameter	27.9MM
Length	39.7MM
3S (LIPO)	1045(PROP)
Max thrust	850G
ESC recommended	18A

It is important to be aware of each of the following specifications:

- The KV value indicates how many RPMs (revolutions per minute) the motor will make when provided with V number of volts. Note that increasing the KV does not mean that your quadcopter will fly better or that it will be more stable. There are many important factors that we should be aware of. For example, more KV means more thrust. As a result, the motors will consume more battery and the flight time will decrease.
- The weight value indicates the weight of the motor. Note that we have four motors in our build, so we have to sum up for four times the weight of each motor.
- The only time when you should worry about the length, shaft, or diameter is when you build your own frame and you have to drill some holes.
- Lastly, the current and voltage will have an effect on the **electronic speed controller** (**ESC**), which will be mentioned later. For now, just keep in mind that we have to choose the appropriate ESC according to the current and voltage.

Choosing a motor

The first thing that we should ask ourselves is about the total weight that we are expecting to lift with our quadcopter. At first, it is difficult to calculate the weight as we have no idea how much it will be; so we usually choose all the parts apart from the motor and estimate a possible total weight.

Since the weight factor must be eliminated by the total thrust the quadcopter produces, it is important to choose a motor that has the desired thrust. Generally, motors should provide a thrust that is two times the weight. This ensures that the quadcopter has the best balance over those factors. Note that it is important to keep the balance. In other words, if the weight is too little or too much, compared to the thrust, the quadcopter will either have extreme sensitivity to all its maneuvers or it will not lift at all, respectively.

Motors that fit your needs can be found in the market. However, it is necessary to keep in mind that more weight, even if your motor can eliminate it, means reduced flight time. After all, it's all about the stability and flight time.

Choosing an appropriate propeller

A quadcopter uses two clockwise and two counterclockwise propellers. They are usually classified by their length and pitch. For example, 10×4.5 propellers are 10 inch long and have a pitch of 4.5. There are two factors that characterize a propeller, dimension and material.

Choosing the right dimension

When choosing a propeller dimension, you need to find a good balance between the length and pitch. For quadcopters that do acrobatics or fly in races, we need acceleration and this means that we need torque. So, you will need propellers with small pitch and high KV motors; so that you have more RPMs and higher torque. As a result, your quadcopter will have more acceleration and eventually will be faster.

On the other hand, for larger quadcopters that carry payloads such as a camera, large propellers and low KV motors tend to be the best choice. These have rotational momentum and easily maintain your aircraft's stability.

Choosing the right material

There are different types of propellers, such as plastic, carbon fiber, and more. Plastic propellers are cheaper and more flexible. This means that there will be more **blade flapping**, which will make your quadcopter more unsteady; this is something that we don't want. However, in the worst case scenario, if you have an accident with that propeller, you will not have a serious injury. Also, when plastic hits an object it may not get damaged. On the other hand, carbon fiber propellers have less blade flapping because they are harder. If a spinning carbon fiber propeller touches your skin, it will be a serious injury, so you may have to protect your propellers somehow.

It's important to mention that blade flapping reduces flight time and craft stability. So, for your overall craft build, it's better to buy carbon fiber propellers and keep a safe distance from any person that may be near it.

Gathering more information

It is a good practice to ask questions in Facebook groups or forums. There are many people who have different builds of quadcopters and the information about the motor and propeller used may be valuable. Furthermore, when buying from eBay or other sites, there are offers along with the motor that recommend you the propellers. Usually, there will be a diagram that explains how much thrust the motor will have as an output when a specific propeller is mounted.

`http://www.robotbirds.com`

Electronic speed controllers

The electronic speed controller or ESC uses a PWM signal coming from the flight controller to create an average voltage/current to control the speed of the motor. There are three sets of wires; one set goes to the batteries, one set goes to the flight controller, and one set goes to the motors. It consists of three wires, which we usually define as servo wires, two of which of are marked as **Vin** (small red) and **Ground** (small black). The third one is the **PWM signal** (white) from the flight controller. The stronger the signal, the more the current through the ESC. It is also connected to the battery via a T connector using a positive and a negative wire. Furthermore, the ESCs output is through the three wires that are all connected to the motor (blue), as shown in the following image.

Lastly, there is another input coming in from the battery. Each ESC has to be connected through a power distribution system to the battery. As shown in the following image, the two left wires (red and black) must be connected to the battery to power up the ESC and the motor. To sum up, ESCs have an input of 12V (depending on the battery used) and they have wires that go to the motor giving an input to the motors from 1V to 12V.

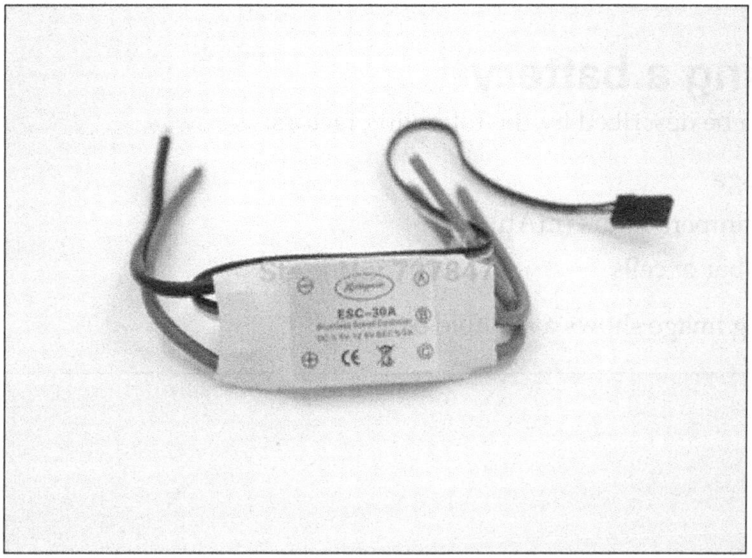

`http://www.aliexpress.com`

As mentioned earlier, we need to have two clockwise and two counterclockwise motors. Therefore, by swapping any two of the three wires that connect the motor and the ESC, we can invert the rotation. This is a pre-flight check and will be described in *Chapter 5, Electronics Installation*. For now, connecting the wires in any combination will be fine.

Choosing the appropriate ESC

Many motors have a recommended ESC. The specification that changes in different ESCs is Ampere. When we covered motors in the *Common motor specifications* section, there was a table that had a recommended ESC of 18A. So according to that motor, the ESC that we buy should be at most 18A.

Battery and charger

Batteries usually last for 10 to 20 minutes. By the end of a flight, you will have to recharge your battery to prepare your quadcopter for another flight. There are various batteries and chargers out there and almost all of them will work. However, when building a quadcopter for aerial photography, we will not just hover over the ground so we need something lightweight.

Choosing a battery

A battery can be described by the following factors:

- Voltage
- Milliampere hours (mAh)
- Number of cells

The following image shows a suitable battery:

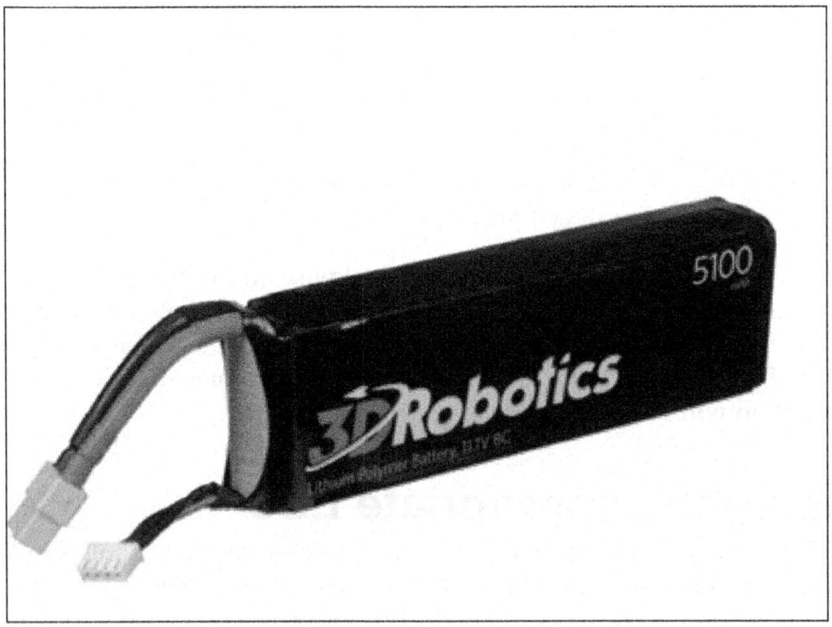

`http://www.camerax.com`

Most DIY quadcopters have a 12.1V battery with 500 to 5000mAh. The total flight time depends on the mAh that your battery has but keep in mind that the higher the mAh the more weight on your quadcopter. So it's crucial to have a good balance and not buy a 10,000 mAh battery that will weigh 800 gms. There are batteries with 2, 3, or 4 cells. You will notice that in specifications the battery will have 3S or 4S. The S rating is the number of cells the battery has. A general rule says, the more the better but it also depends on your charger and on your ESCs.

So to conclude, a good choice should be a battery over 2500mAh but lower than 3500mAh with 12.1V and 3S.

Choosing a compatible charger

The batteries last for 10 to 20 minutes. After that you will need to charge your battery and prepare your craft for another flight. It's crucial to reach over 15% of your total mAh before charging. Many chargers will notify you with a message of low voltage. There are chargers that charge only LiPo batteries and chargers that charge 2 or 3 different battery types; it really depends on your budget. However, many people say that a good charger will make your battery last longer and keep your craft from voltage spikes, which may cause a crash.

http://www.skyrc.com

An IMAX-B6 charger can charge LiPo and other types of batteries. Usually, multicopters use LiPo batteries but there are cases when you might need something else. It's always a good choice to have something compatible with NiHM batteries too.

Power distribution boards

Usually, quadcopters have one battery. But how can we connect one battery to four ESCs? A power distribution board is the answer. It is a simple board that has two pins as input to the board and four or more outputs. Therefore, we can connect our battery and have up to 8 outputs and each one of them is connected to the battery. The following image is of a simple power distribution board that can be found in the market:

`http://quadcopters.co.uk`

As you can see, there are four blocks of two outputs. Each output consists of two pins: one positive and one negative. There are two more pins, one positive and one negative, for your battery connection.

Make sure that this board does not touch any other conductive material. To conclude we can say that through this board we can power up all our ESCs with the voltage of our battery.

Radio transmitter and receiver

Remotely piloted quadcopters are controlled from a ground station using some kind of transmitter and receiver mounted on the craft. This communication system is simple, reliable, and easy to install. Note that the ground station is actually a pilot with a transmitter in his hands. A ground station can have many more systems but a simplified model of communication is preferred. Transmitters usually look similar to the following screenshot:

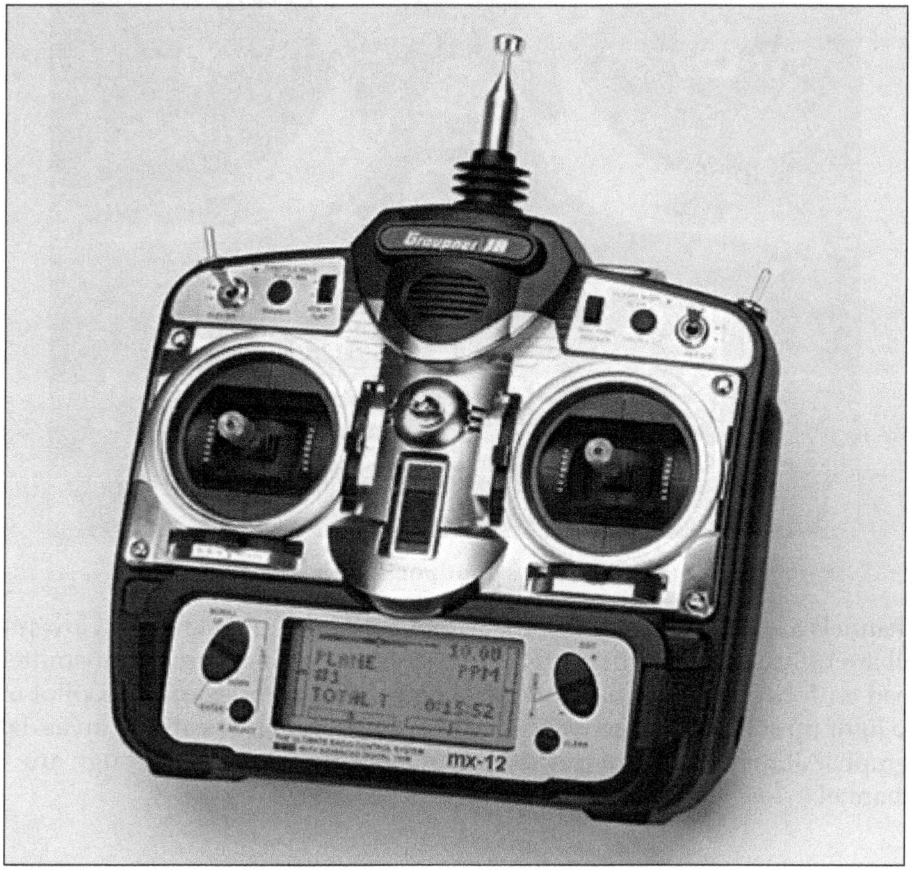

http://www.quadrocopter.com/

They have two sticks and a screen for all the configuration setups. A joystick can also be used as a transmitter but a proper transmitter is recommended for new pilots. The flight controller that will be described in later chapters has all the configurations needed in case we want to connect a joystick or a transmitter similar to the one shown in the preceding image. Let's have a close look at what a transmitter consists of:

`http://www.quadrocopter.com/`

Four channels are used, namely throttle, rudder, aileron, and elevator; yaw, pitch, and roll are fully controlled through these channels. Furthermore, a transmitter is equipped with two or more channels for extra functions; for example, a pilot may need to light up some LEDs, activate a system while flying, or capture an aerial photograph. All these are extra functions that a pilot can activate through any extra channel.

http://www.quadrocopter.com/

Usually, these systems have some kind of memory that stores its data even if the battery is disconnected. Memory is required to save all the configuration a model has. For example, let's say that we have two airplanes, one helicopter, and one multicopter and we need to save all the configuration of our models without any conflict. The transmitters have this option and we can try to fly our craft, change the settings, fly again, modify configurations, and then lock our configurations and save them for future flights.

Apart from the four sticks that any pilot uses to control its craft, there should be four trim sticks with which we can change the trim of the craft according to the drag it has. Assuming that the minimum throttle is 1000 and the maximum is 2000, trimming the throttle channel has an effect on the min-max range. In other words, if we trim the throttle, the minimum may be 1050, 1100 or even 1200. Practically, at the exact moment that we pull our stick a bit, the minimum throttle will be sent to the flight controller. More information can be found in the later chapters.

For each transmitter, there is a receiver of the same GHz. The receiver simply receives a signal from the specific transmitter and through the right pins passes the signals to the flight controller, as shown in the following image:

http://www.amazon.com

Quadcopters can be programmed and controlled in many different ways. Usually, there are two modes: acrobatic and stable mode. The difference is the way the controller board interprets the orientations feedback, together with your RC transmitter joysticks.

As shown in the following image, this is the transmitter that Parrot uses in some of their quadcopters. Even if it is quite strange, it actually consists of two sticks and a tablet where you can see what your quadcopter sees:

http://droneflyers.wordpress.com/

Summary

By now, almost all of the parts needed for your build should be clear and you should have a basic understanding of each part's usage in the overall build. Before mounting everything on our frame, we need to solder and prepare all these and some more electronic parts and of course create a frame, as we will see in the next chapter. More information about the connections will be provided in *Chapter 5, Electronics Installation*.

3

Creating a Frame

Even the most skilled pilot will find it hard to fly a quadcopter built over a badly designed frame. The frame is the base of the quadcopter and one of the most important parts. It is something that cannot be changed easily. Everything is mounted on the frame; thus, removing and changing any component will be very time consuming. It will require removing and remounting of all the components. There are various shapes and materials with which we can build a frame and every single one of them has its own advantages and drawbacks, which we will explain later in this chapter. A good and well-built frame ensures a stable and easy flight. On the other hand, a frame that is built quickly with little attention to detail can cause stability problems, drag, and may cause crashes. Since safety is the number one priority, you should be cautious when building your own frame.

In this chapter, we will cover the following topics:

- The market to buy frames
- How to build a DIY frame
- What material to use and why
- In which shape should you mount your electronics

Market versus DIY

After years of testing and designing, many companies including DJI and Parrot have created some awesome drones. Inspire 1 is currently the best drone from DJI and Parrot quadcopters are made on some very beautiful, light, and stable frames. However, those frames are not customizable. Mounting your own electronics on a market frame may not be possible or may require too many modifications for it to be practical.

Furthermore, there are many simple and very good frames available on websites like `http://ebay.com`, `http://amazon.com`, or even `http://hobbyking.com`. The cost depends on the material and the number of motors that you want to mount. Buying a frame from websites like the ones mentioned is an excellent choice; if you don't want to mess up by designing and creating your own frame using any kind of material.

In the left-hand side image (DJI HJ450 frame), the frame is made from plastic and in the right-hand side image (X450), the material used is aluminium:

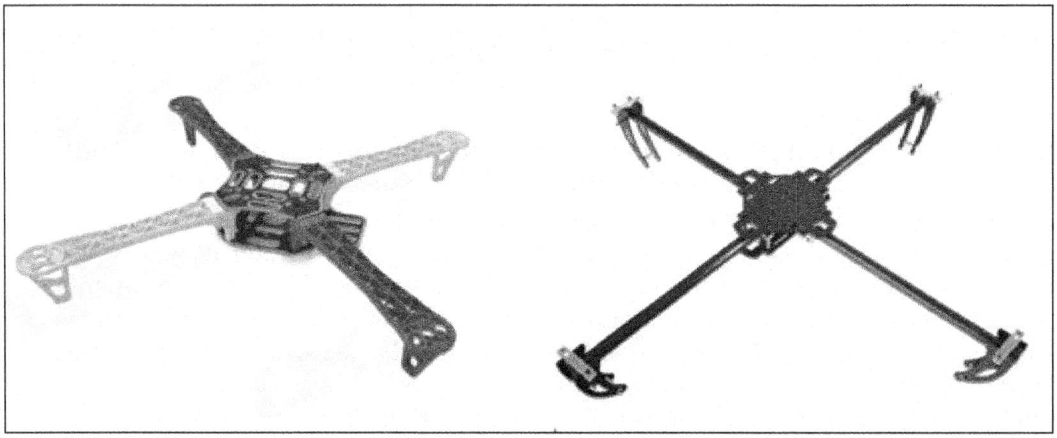

The following image shows two more frames made from aluminium and carbon fiber:

As a third option, we can create our own frame step-by-step. You should keep in mind that a wooden frame will work just fine and your quadcopter will fly; however, it is not the best thing you can do. Since the first decision we need to take is what material we are going to use, let's have a look at all the possible materials.

Materials

Depending on the budget, various materials can be used to build your own frame or you can buy a market one. A frame does not need to be only one material. For example, you will notice that many frames consist of a combination of different materials.

Wood

Wooden frames are simple, easy to build, and more importantly, you can make any change you may think of, later on. As far as the frame is concerned, the cost of damage after a crash is very low. These are almost all the advantages of a wooden frame. You can either buy some wood and cut it by yourself or give instructions to your local shop about the dimensions. Lastly, when creating a wooden frame, make sure to cut it with a laser; this way you will be able to design it precisely according to your dimensions and hopefully have a stable quadcopter.

Aluminium

Aluminium is the second and most common material used for quadcopters. It is quite simple and lightweight and suitable for hobbyists who seek fun. Aluminium is lighter than wood and is much more stable. If all the screws and electronics are mounted on the frame with caution, the overall build will be nothing less than a stable, lightweight, and reliable quadcopter. Assuming that we are comparing the materials, a drawback of using aluminium frame could be the weight. Although it's a very good material, there is always something better and more expensive, for example carbon fiber.

Carbon fiber

Carbon fiber is by far the best material for crafts. It is light but not so much that it increases the vibrations. It's hard, so it usually prevents your quadcopter from getting damaged after a crash; it is complicated too. Assuming that you need to build a DIY frame with your own design, this material is not so easy to cut. It is also expensive in relation to the other materials.

Plastic

The last choice is plastic. Plastic is used as a material for frames coming from a manufacturer but most importantly, plastic can be used in 3D printers to design and finally print your own unique frame. 3D printers are an amazing breakthrough, they are small and efficient objects because of their low cost, build, and easy design. Many software, such as Blender, Rhinoceros, and SketchUp can be used to export your own design as a file.

http://www.gigaom.com

After that, the file can be imported to the printer's system and finally you can view your design. However, as every material has its own drawbacks, 3D printing is quite hard for beginners. The main reason is because it requires some basic knowledge when creating a design and when setting up the parameters, such as the density of the frame or the dimensions. The details matter here and creating a frame without caution may lead to a crash.

Conclusion

After going through almost all of the main materials that can be used to build a quadcopter, it is important to have an overall view of all of them. Wood is the best material for cheap and easy to build quadcopters, aluminium is a mediocre choice that provides stability and is easy to build too. Carbon fiber is the best material as far as the vibrations and the after-crash results are concerned, but it's quite expensive. Lastly, plastic is something that can be used to 3D print your own design with accuracy. However, there are many factors that determine if your craft will live another day after a crash and how much weight it will have. Choose your material wisely!

Now, let's have a look at the possible shapes.

Frame shapes

There are plenty of shapes out there and all of them are just fine. There is no big difference in the shape, as long as it is created with caution. Two of the most common shapes were mentioned in *Chapter 1, Introduction to Quadcopters*. Apart from the X and plus shapes, there is an H frame and other more complicated ones. Generally, most hobbyists choose the frame shape according to what they believe will be more beautiful and elegant. Furthermore, there are some techniques and shapes that are suitable for fast quadcopters, in case you need as much acceleration as you can get. There are not many races out there but, in the future, races using quadcopters are going to be very common and everyone will try to get as much acceleration as they can. Note that the discussion in this section of the chapter will be about shapes and not sizes.

The X shape

We already mentioned that this kind of shape is the most common of all shapes. The X shape defines the basic shape and it is usually the first attempt to build a quadcopter. The reason that it's defined as X is because the quadcopter looks like the Greek letter X from the top view. Each motor is mounted on the edges of the shape and everything else is usually placed at the center, as shown in the following figure:

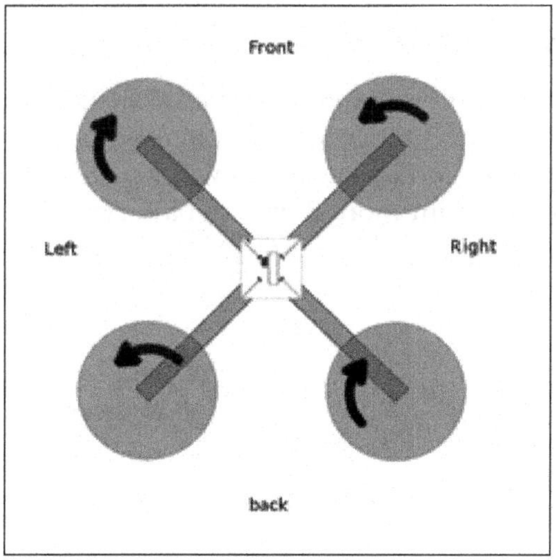

In comparison to other shapes, this shape is advantageous because the total thrust needed for acceleration is split in two halves, since two motors will be used. This also increases our motor efficiency.

The plus shape

The plus shape is exactly like the X shape with one difference; moving forward will require one motor and not two, as is the case in the X shape. Also, with a plus shape it's quite difficult to mount a camera at the front because the front propeller is in the way:

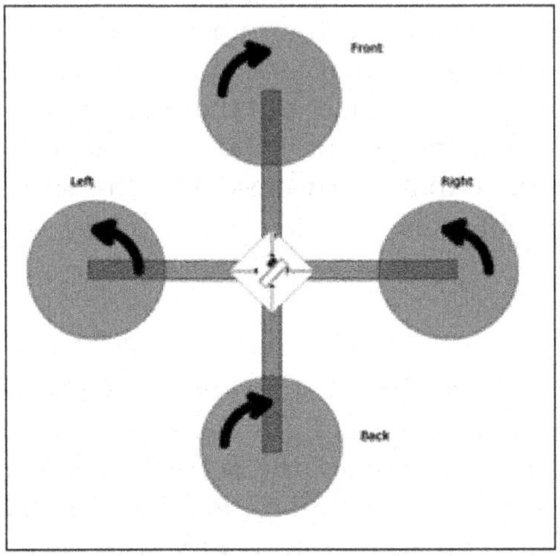

The H shape

Like the Greek letter H, with this shape we can add more electronics, cameras, and DIY stuff that we want on our quadcopter. The main difference between the X and plus shapes is the space between the arms. As we can see in the following image, there is plenty of space to mount more than one battery:

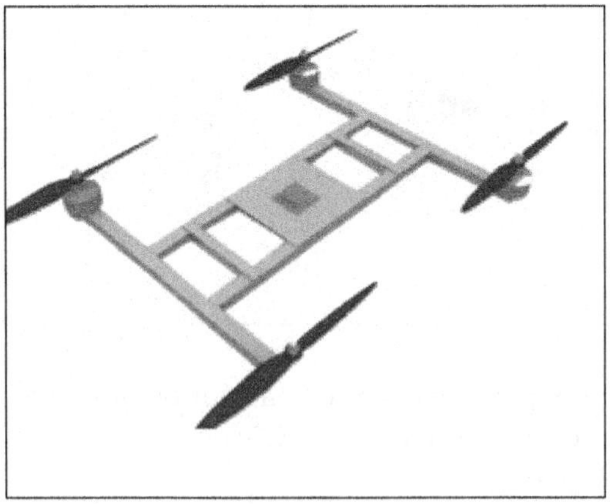

http://stigern.net/

The racing shape

Quadcopters intended for a race need to have a specific shape. The two front arms must be lesser in length than the two at the back. This increases the acceleration and the quadcopter gains more speed. Furthermore, hobbyists usually try to build as small a quadcopter as possible. All the electronics are placed in the center of the total weight, in other words, around the center of gravity.

The stability shape

Quadcopters made for aerial photographs and video capturing must have a good stable frame to minimize vibrations. Usually, frames for such a purpose are big. The four arms have the same length and are very long to increase the stability of the overall system, as compared to the racing shape where all the electronics are placed in the center of the total weight.

Summary

To sum up, there are many different options to choose your shape and the material from. Keep in mind that almost everything will work just fine, but since the purpose here is to build a quadcopter for aerial photographs, the lighter it is, the better result you will have. The best combination on shapes and materials is to use PLA plastic with carbon fiber arms so that your quadcopter will be light and stable too. Since the frame is ready, we can now go through the details about mounting our electronics on the frame.

4
Soldering the Electronics

Most components need extra modifications before connecting them together. Electronic speed controllers have to be soldered to the connectors, to make sure that the overall build is highly customizable for future plans. Note that there is always an option of not using any connector, which will be described later; but if someday you need to modify or add something, you will have to heat the connections and install a new connector, which is not recommended.

Each component has numerous wires with different connectors for different purposes. These connectors allow for easy swapping or upgrading of components, without the need to disorder the wires, which is time consuming. Most components come with standard connectors that are already soldered, while others may require you to separately purchase and solder your own connectors. This chapter will describe what connector is used in what situation and where you can purchase and solder your own connectors.

This chapter provides information about the following topics:

- Information on motors
- ESC
- Soldering a banana connector
- Soldering a T connector
- Soldering a power distribution board

Bullet connectors

Bullet connectors are going to be used to connect the motors with the ESCs and the T connectors are going to connect the ESCs with the PDB.

It's extremely important to place the correct connector in each component. In every component that is already connected or produces power, we must always use a female connector. We don't want to expose any kind of metal that is connected to power; with this said, we can make the following connections:

- Female banana bullet connectors to the 3 wires coming from the motors
- Male banana bullets connectors on the 3 wires coming from the ESC
- Female T connectors to the wires coming from the PDB
- Male T connectors to the wires coming from the ESC

Further in this chapter, we will use the following connectors:

- 12 pairs of 3.5mm bullet connectors
- 5 pairs of T connectors

Motors

Motors have only three wires coming from their insides. Usually, all three are connected on 3.5mm diameter banana connectors. Obviously, in the worst case scenario, we will have to solder these connectors. If you have no banana connectors, you should consider buying some, as they are quite easy to install and are reliable. The following image shows wires with and without the banana connector. At the end of the chapter, there will be a short guide and instructions on how to solder a connection like this:

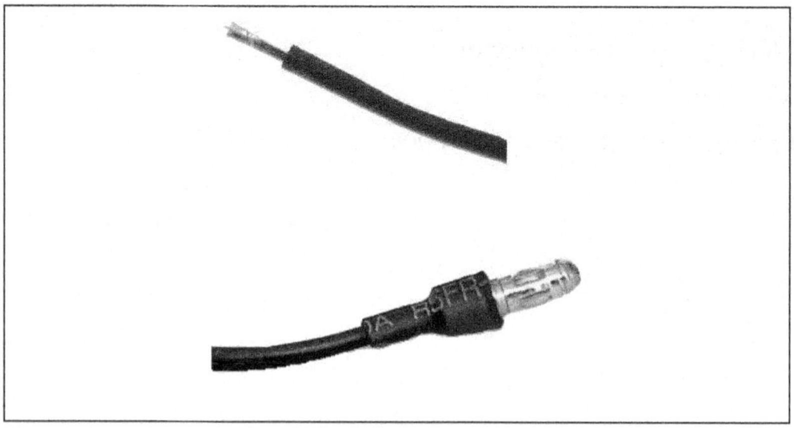

ESC

An ESC has three wires that must be connected to the motor, two wires for the battery and a servo wire that goes all the way to the flight controller. Unfortunately, almost all of the ESCs come with no connectors. So, it is up to you to do all the work and build up the connections.

As you can see in the following image, the three wires and the connectors have been covered with heat shrink. The reason we are using a heat shrink is to protect the exposed metal of the connector from electrical shorts:

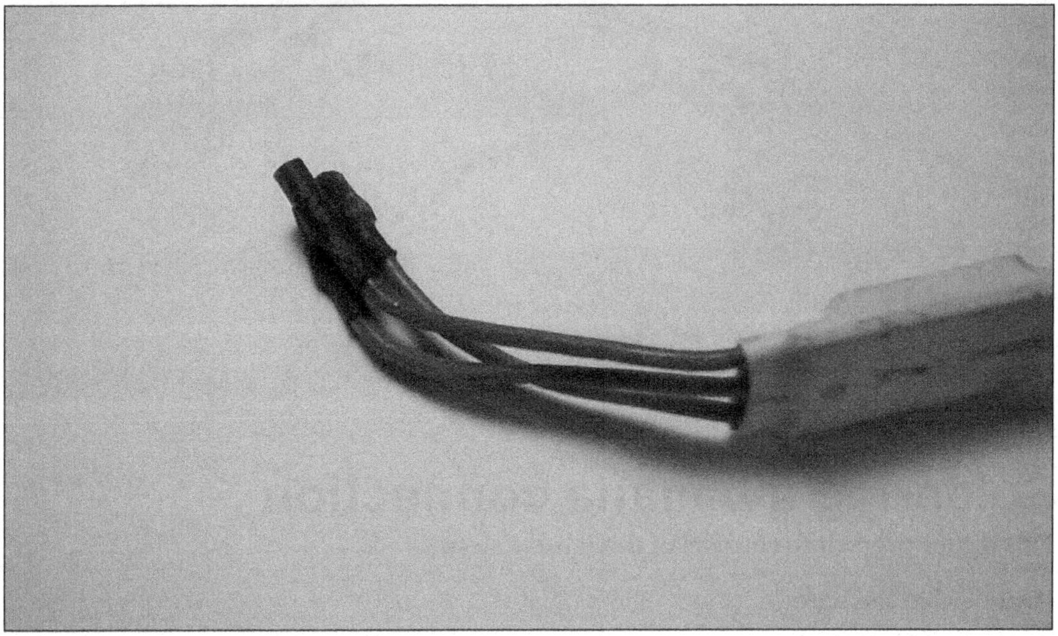

On the other side of the ESC, the black and red wire must be connected to the battery through the power distribution board. Again, this step is optional and we can solder the two wires directly to the power distribution board, but if we need to modify anything or change our frame later, we will have to remove these connections and heat the solder to disconnect the wires.

The following image depicts the ESC after soldering a T male connector. Note that a female T connector should already be soldered to the power distribution board so that the ESC can be powered on:

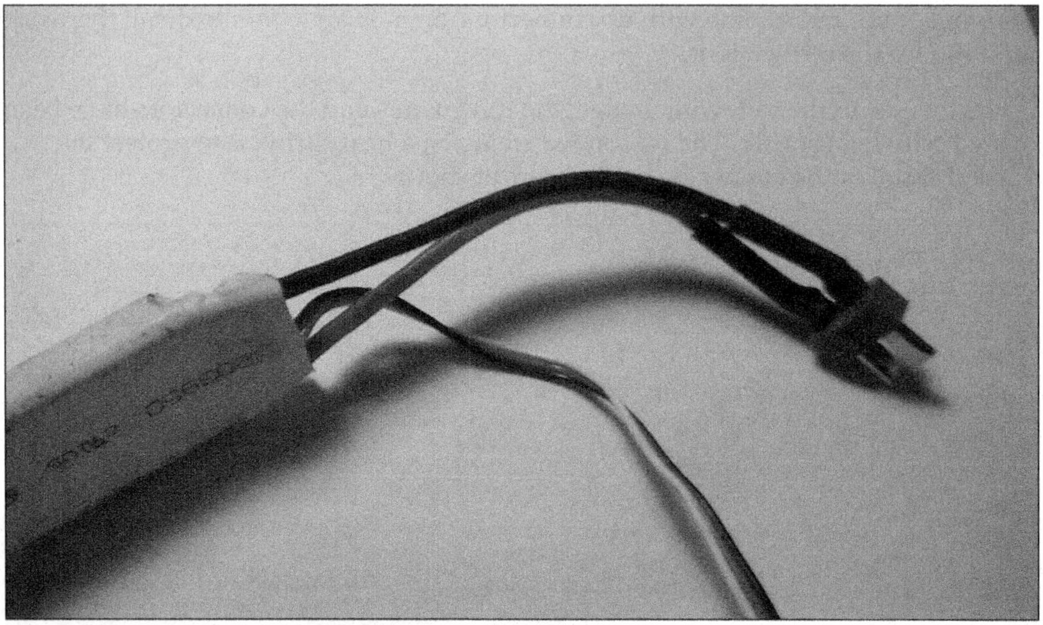

Soldering a banana connection

This solder procedure consists of three basic steps:

1. Strip the wire
2. Solder the wire
3. Heat the heat shrink

Stripping the wire

There are many tools that can help you to strip wires. You can buy them at your local hardware store or Internet markets like eBay. Then place your tool about 1 cm from the end of the wire and remove the cover; depending on the wire, you may be able to gently cut your wire with scissors and remove it with your hands:

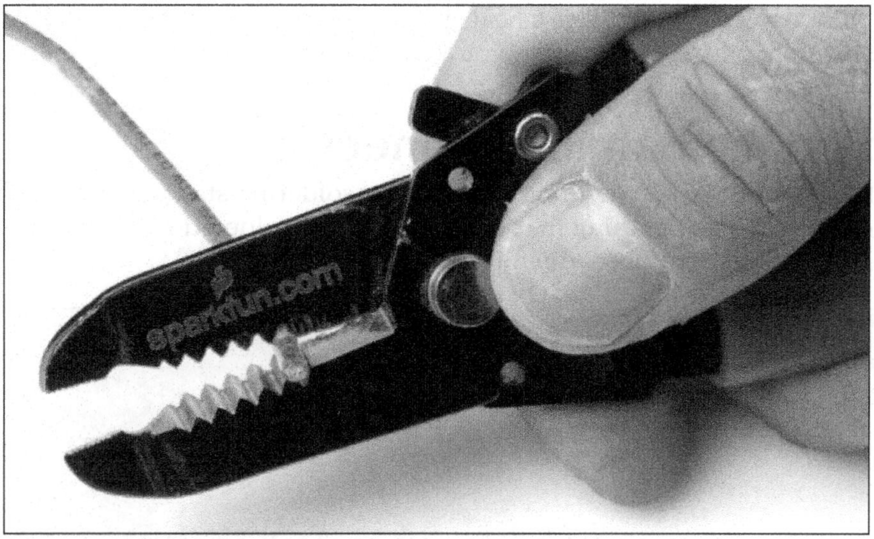

http://www.learn.sparkfun.com

After stripping, your wire will be ready for soldering:

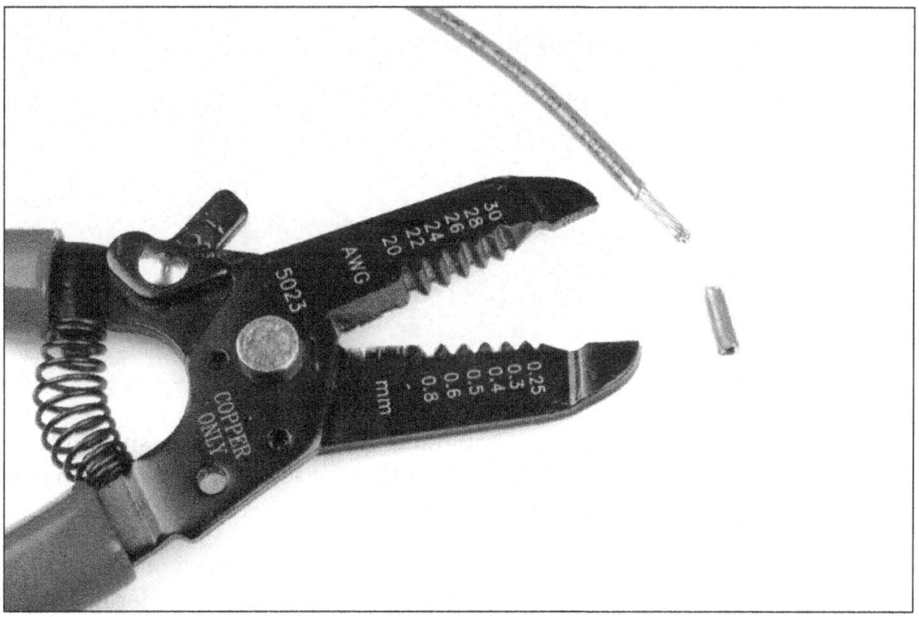

http://www.learn.sparkfun.com

At this point don't forget to place a heat shrink over the wire because once you connect the wire with the connector, you will not be able to do that.

Soldering a banana connector

Be careful when soldering anything. Almost any soldering station has enough heat to injure your skin. So anything that you are trying to solder, do it with caution and with a clear table. Place your wire in the paste, as shown:

Next, we need to place this wire into a banana connector. The wire should be tinned by heating and applying the solder, but make sure not to clump it. It also has to be hot enough. When you do it right, the solder sucks up the wire leads.

The connector will become very hot; so you should hold it with some kind of "helping hand". If you don't have any helping hand you may consider asking for help from a friend. He can hold the wire using any kind of grabber mechanism. Make sure that you need to keep the wire steady using whatever you can. Lastly, you should use a cardboard. The solder often drips and cardboard is a cheap and replaceable surface to protect your workbench/table:

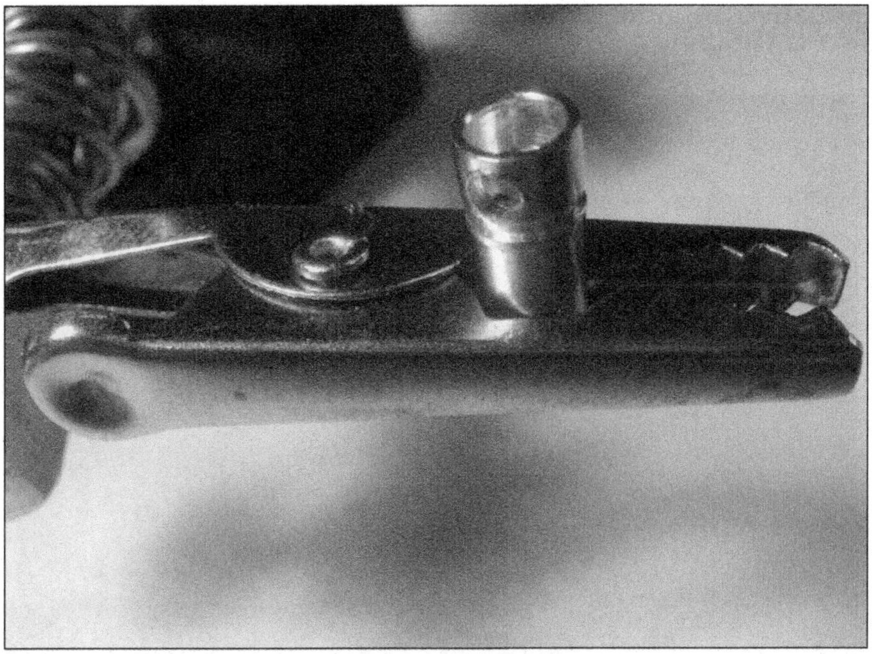

Proceed with soldering the connector, as seen in the following image:

Finally, place the wire inside the connector. After you place the wire in the banana connector, keep it steady for about 8 seconds:

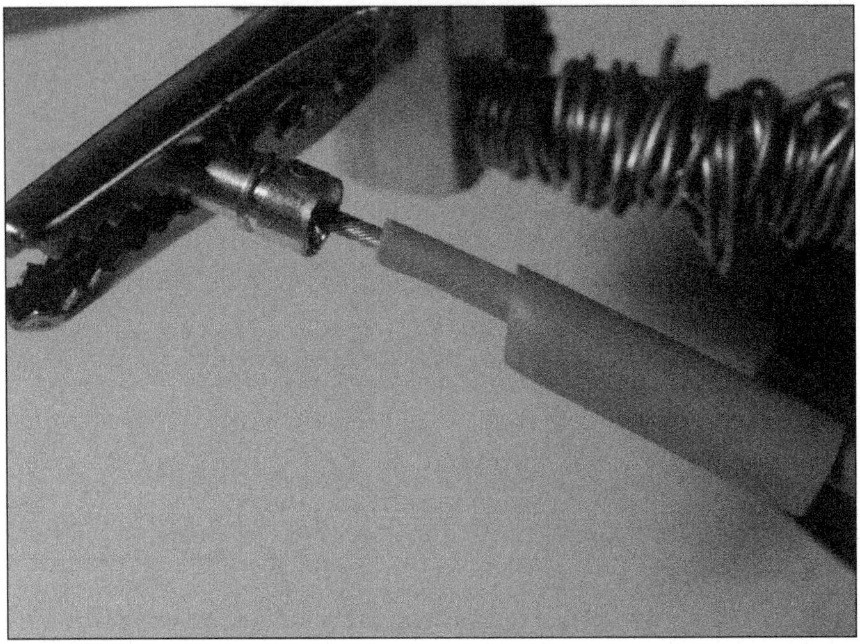

After that, just heat the heat shrink to the right position. Note that I had to use a slightly larger heat shrink because the wire could not fit in the connector:

After heating the result should be the same as shown in the following image:

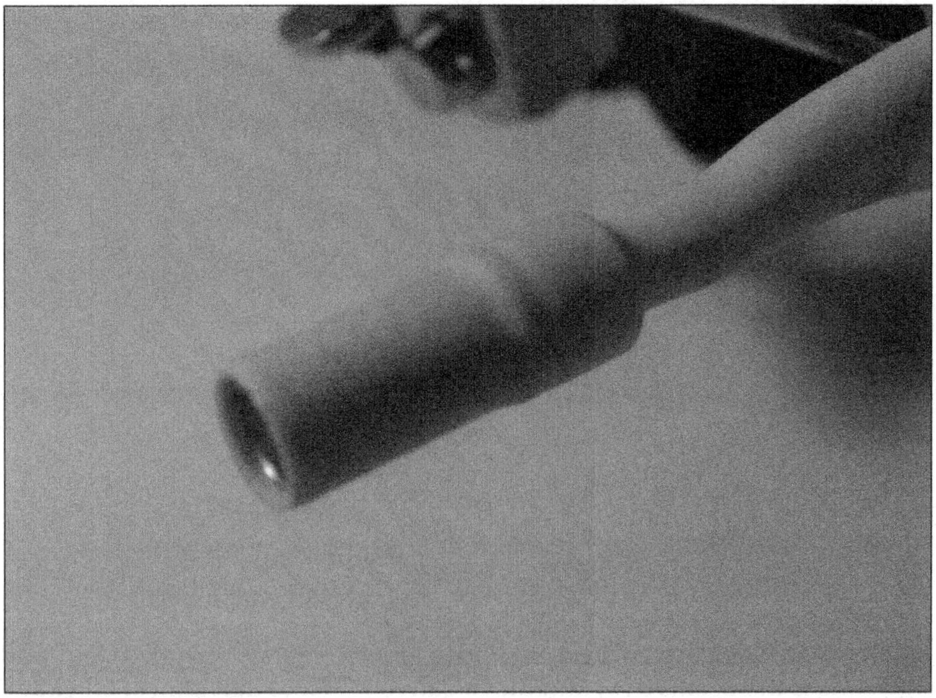

The female heat shrink should be completely covered as shown in the preceding image. The male half should be covered till the point that it indicates how much of it goes inside the female.

To verify if your connection is good, try to pull apart the wire with the banana connector gently. If you can remove this easily, then repeat all the steps; you will have to solder your first connection correctly. Remember that you need to solder two connections (male and female).

Soldering a T connector

Soldering a T connector is as simple as a banana connector. The only difference is that instead of one you will have to solder two pins; one for the positive and one for the negative wire. This kind of connector is called T because it consists of two pins in the shape of the Greek letter T. It is quite useful for these kinds of connections, since it can safely prevent reversing the connector, which will damage the battery and electrical components.

In the following image, the connector is half connected (male to female). At the left and the right side, we will solder our wires:

First of all, we need to strip two wires (if they are not already stripped) and find out which one is the positive (+) and the negative (-). Note that almost all batteries already have a T connector and you can easily find out the positive and the negative pin. Be careful not to solder it in the wrong order as this will destroy your battery:

Hold your connector with your helping hand and lace some solder over the pins. as you can see in the following image:

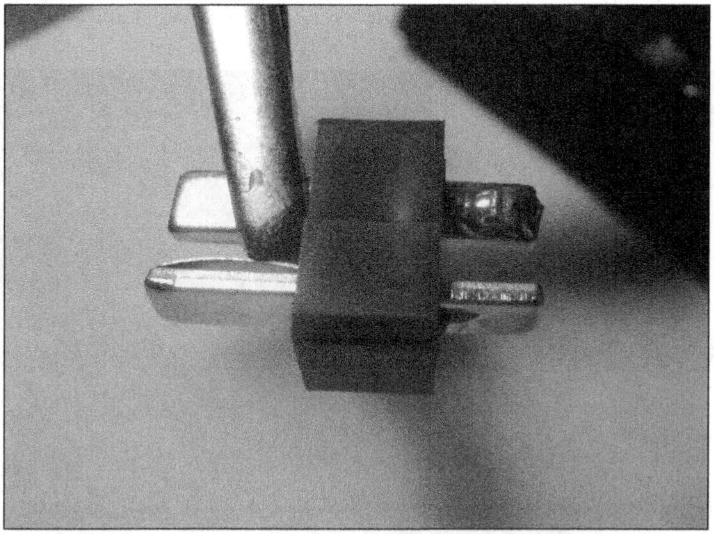

After that, place your wire into the paste as shown in the image in the *Soldering a banana connection* section.

Insert a heat shrink, for later use. Next, gently place the wire over the pin and solder them together:

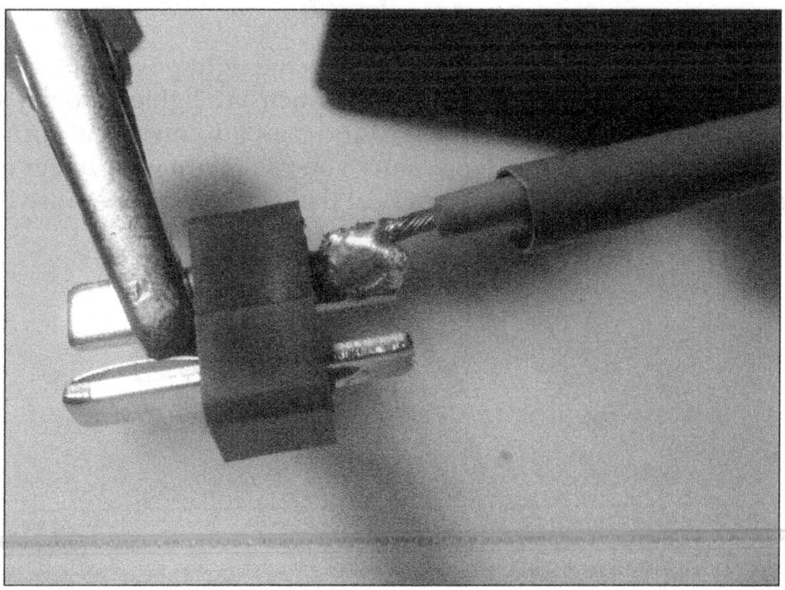

Place the heat shrink to the right. If for some reason it cannot be placed over the pin, choose a bigger heat shrink or repeat the step and solder the wire again and try to place as little soldering wire as possible so that the heat shrink can fit over the pin and the wire:

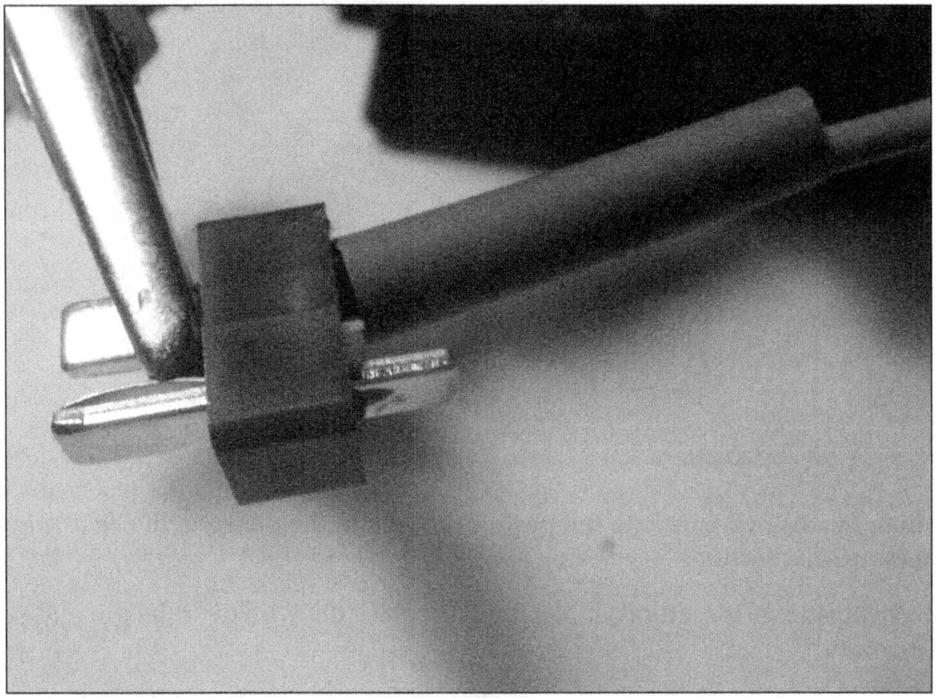

Once everything is ready and the heat shrink is in the right position, try and heat the heat shrink with any kind of heating device, such as a lighter. Do not stay over a specific point for more than 2 seconds. Try and pass the entire heat shrink 5 to 6 times, slowly. Be careful with the flame, since you can injure your finger or burn the plastic part in the T connector. Do not heat the wire for long because it will get damaged:

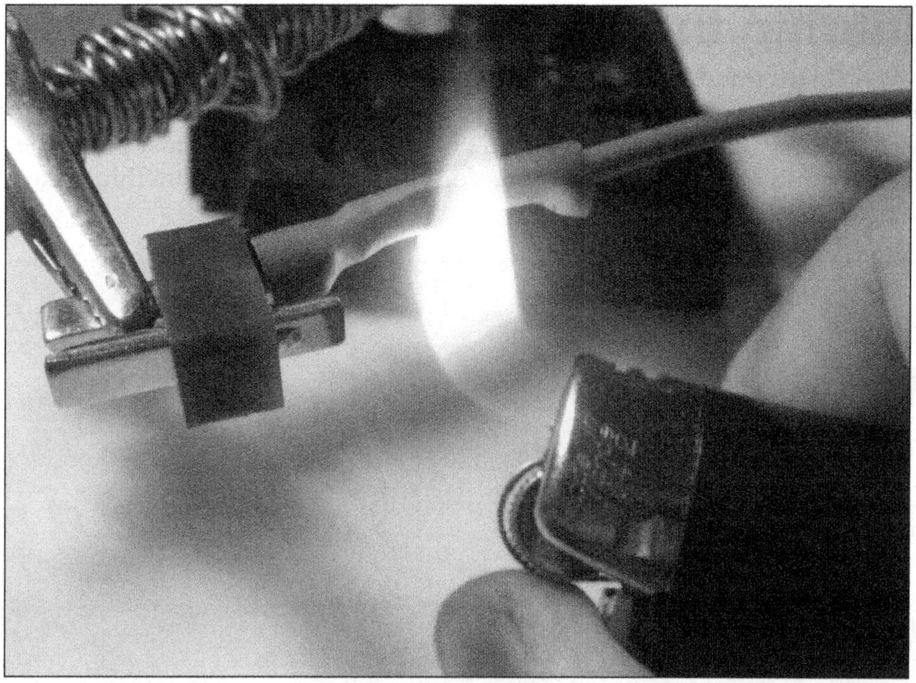

The result should be something like this:

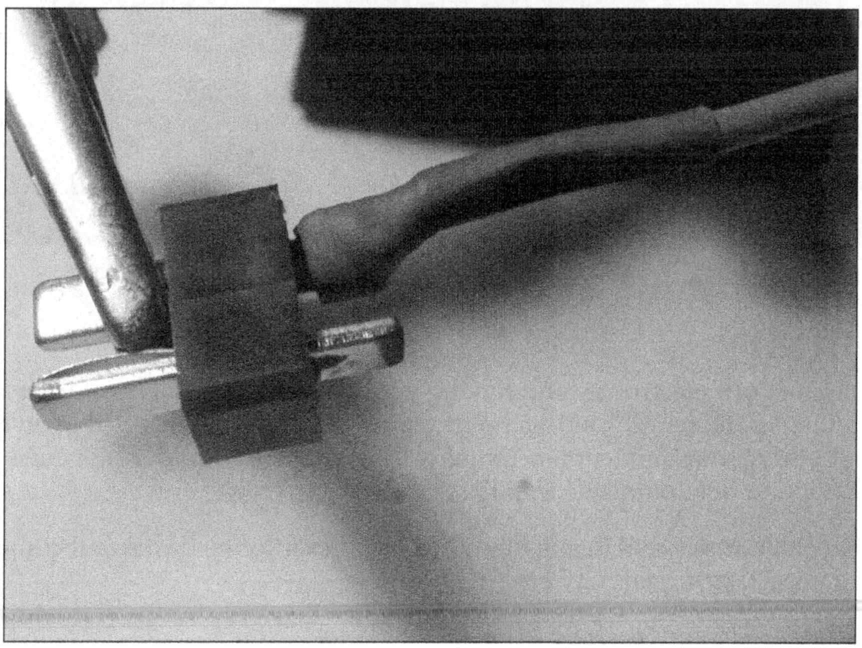

Soldering the PDB

The power distribution board is used to connect all your ESCs to your battery. Every single ESC will be connected to a T connector; so we need another T connector mounted on the PDB. Note that you don't need any heat shrink, just put some paste on your wire and solder it to your PDB. Your setup should be something like the following image:

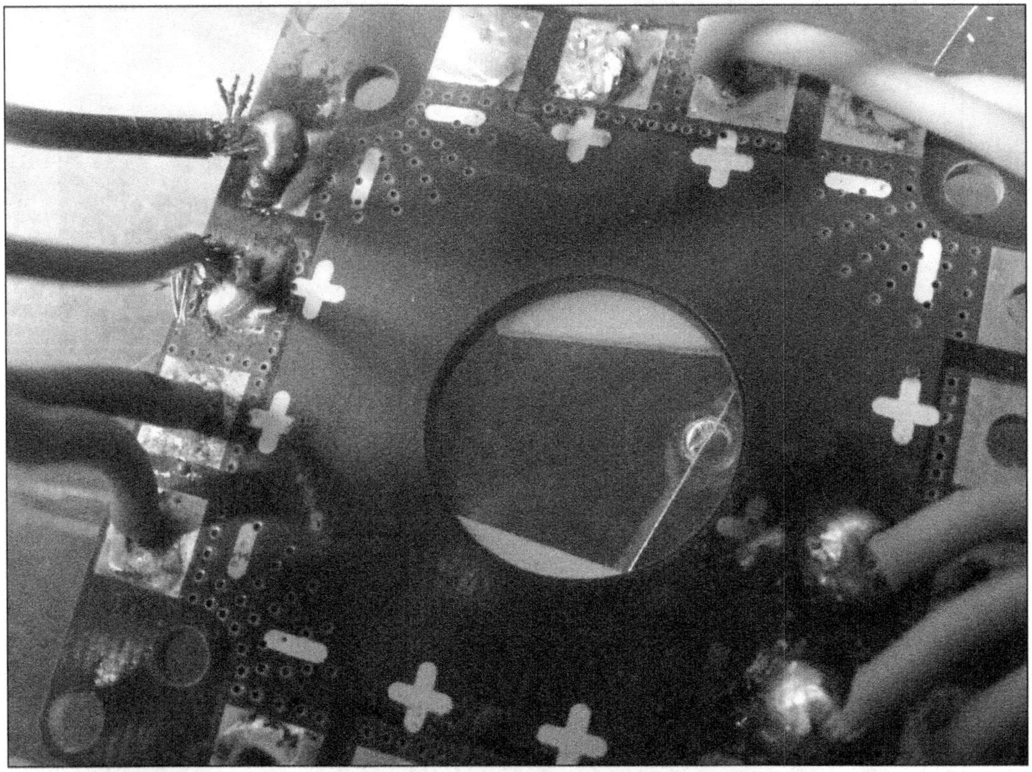

Summary

Taking care of your electronics and having good connections is a must for a well built quadcopter. Hopefully, at this point we are ready to get everything on a table and start building our quadcopter. Usually, after a crash, re-soldering connections like we saw in this chapter is necessary.

In the next chapter, we will install all our electronics into the frame and come closer to building our quadcopter.

5 Electronics Installation

In the previous chapter, we went through the soldering procedure for some of the electronics that needed further modifications. Hopefully, at this point all our electronics are ready for installation. In this chapter, we will mount the motors, connect the ESCs to the motors and the PDB, connect the receiver to the flight controller, and finally power up everything with our battery. Note that all the instructions that follow are crucial for your first flight. Read every step with caution and if the result is not what you expected it to be, please be patient and repeat the step.

In this chapter, we will cover the following topics:

- Motor installation
- Mounting the propellers
- Installing the ESCs
- Powering up

Motor installation

Depending on your motor, this step may be either simple or very complicated. Generally, you need to make two or four holes at the edge of your frame. That *or* has a reason to exist.

Some market frames look like this:

`http://www.builtdrones.com/`

These have all the necessary holes that you may need, assuming that your motor has normal dimensions. So, there is nothing more to do than taking a screwdriver and placing some screws over the four motors.

But if you try to mount your motors in wood, aluminium, or some other material that has no holes, you will have to do some more work. Almost all the motors have some specifications on the market from which you have bought them, as shown in the following figure:

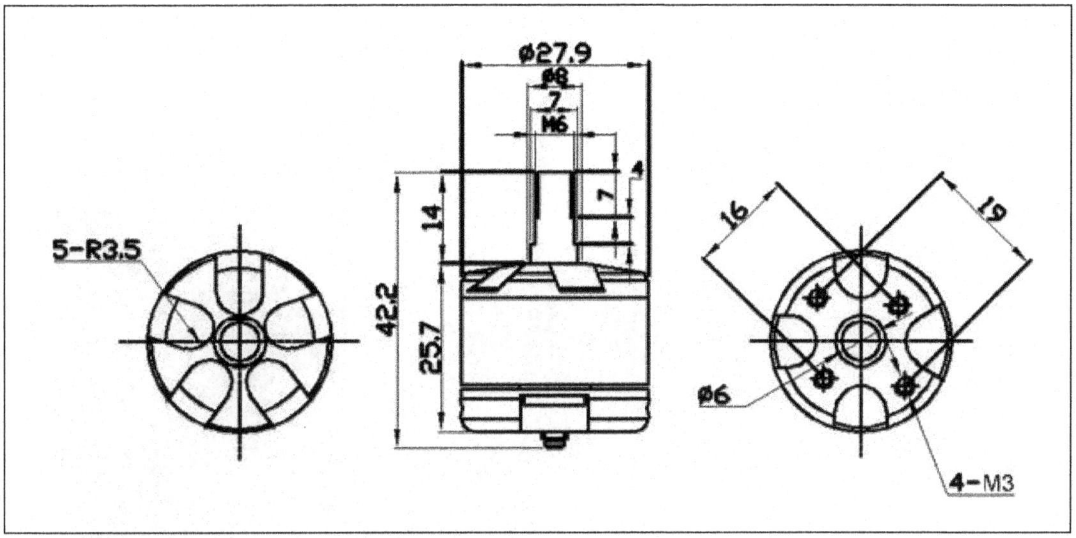

http://www.emaxmodel.com

So, hopefully, you can find the dimensions of the motors and the exact size and spacing of the holes needed. Either you can make some holes with appropriate tools or buy an external motor mount system, as shown in the following image. The motors usually have their own screw thread type and you should check that before buying them. It may be useful to buy some screws, but be careful not to buy a long one and damage the inner coils of the motors.

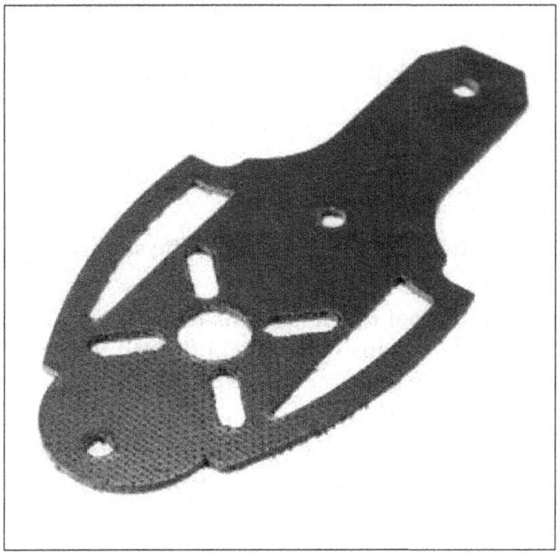

http://www.passion-drone.com

You may have to make some holes for the mount system. It really depends on the system that you buy. Another hack that I have done on my quadcopter is shown in the following image:

It really works. Choose wisely what you will do because it is very important to have a steady and well mounted motor.

Further tips

The motor must be mounted vertically. The slightest angle might cause flight problems.

All the screws must not reach the coils inside the motor because they will damage it and you will lose flight stability and power. If it does, remove it and place another screw or more washers.

If, for any reason, you open your motor, say for cleaning it, secure the safety of the motor by running a test. Spin your motor to full throttle without any blades. Just spin the motor and watch if the safety is on.

Mounting the propeller

Before mounting the propellers in the correct manner, we need to test each one and determine if they need further balancing. Balancing the propellers is important and an unbalanced propeller results in "shaky" hovering of the quadcopter. To minimize the shaking, you have to constantly fix each drag with the sticks on your remote controller and this is quite annoying; especially when the goal is to capture a clear photograph.

There are tools that can be used to balance your propellers easily, as shown in the following image:

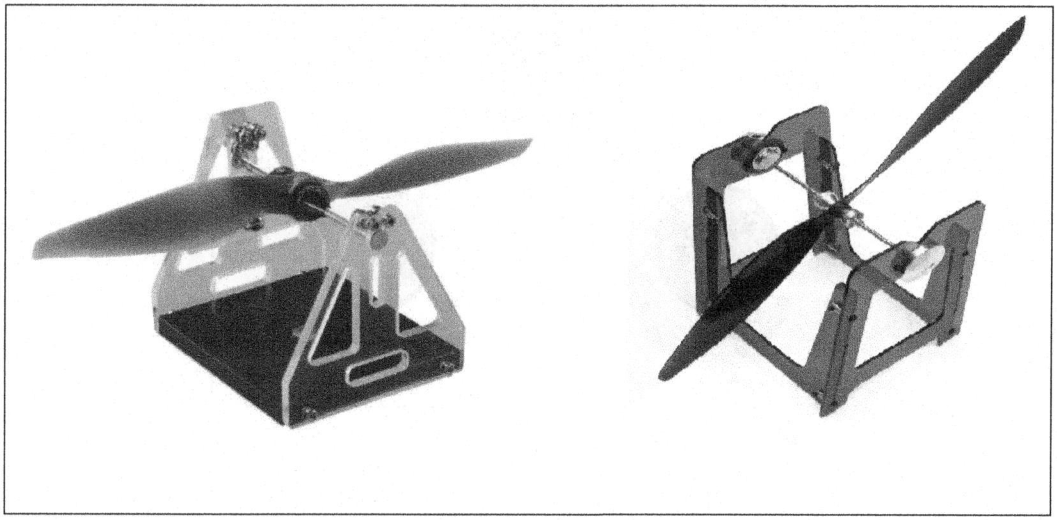

`http://www.builtdrones.com/`

As you can see, the left propeller is perfectly balanced and the right one is not. But how can we fix a badly balanced propeller? We can simply put some low weight on one side of the propeller. For example, you can use Scotch tape. The weight of a 1x2 cm piece is enough to change the balance. Place the tape on the side that is higher and you will see that the extra weight that we have placed balances the propeller perfectly.

There are other ways balancing your propeller, such as sandpaper, but you have to be very careful with it.

Now that the propellers are balanced, we have to mount them on our motors. If you take a closer look at a propeller, you will see that you have two clockwise and two counterclockwise. You should remember the figure from the *Flight dynamics* section in *Chapter 1, Introduction to Quadcopters,* which shows how the motors should spin.

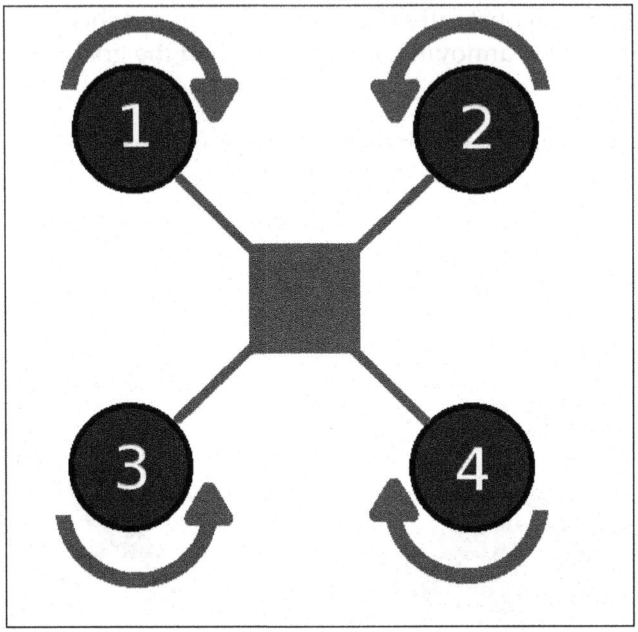

Motor 1 should have a clockwise propeller, motor 2 should have a counterclockwise propeller, and motor 3 and 4 should have a clockwise and counterclockwise propeller respectively.

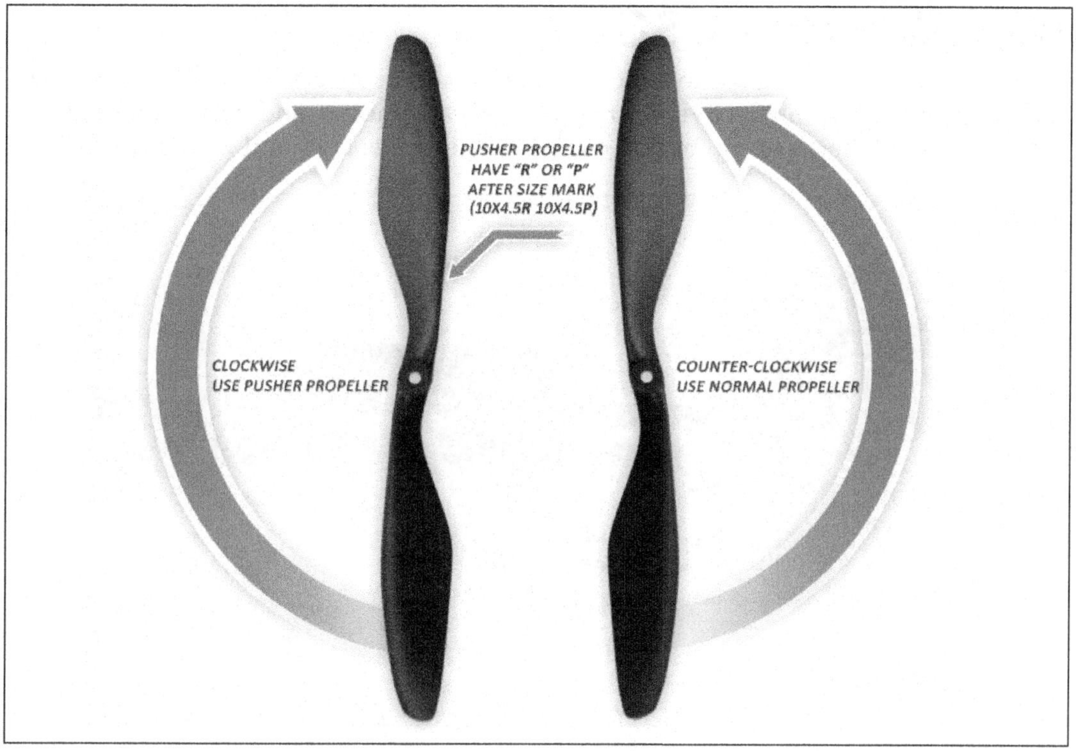

http://www.rcgroups.com

There are two clockwise and two counterclockwise propellers. At the opposite motors, we need to use the same propellers. As a result, front-left and back-right will be the right direction propellers and the other two propellers would go to the other two motors.

Note that every propeller screw is different from a normal screw. You have to screw it in the opposite direction because high RPM may loosen the screw. However, there are many motors that have right direction screws, which is also fine, if they are screwed tight enough.

Installing the ESC

Each ESC must be connected to a motor, the flight controller, and the PDB. Connect the three wires of each ESC to the three wires of each motor randomly. You might have connected them wrong, but it's ok. We will test this later and fix it if needed.

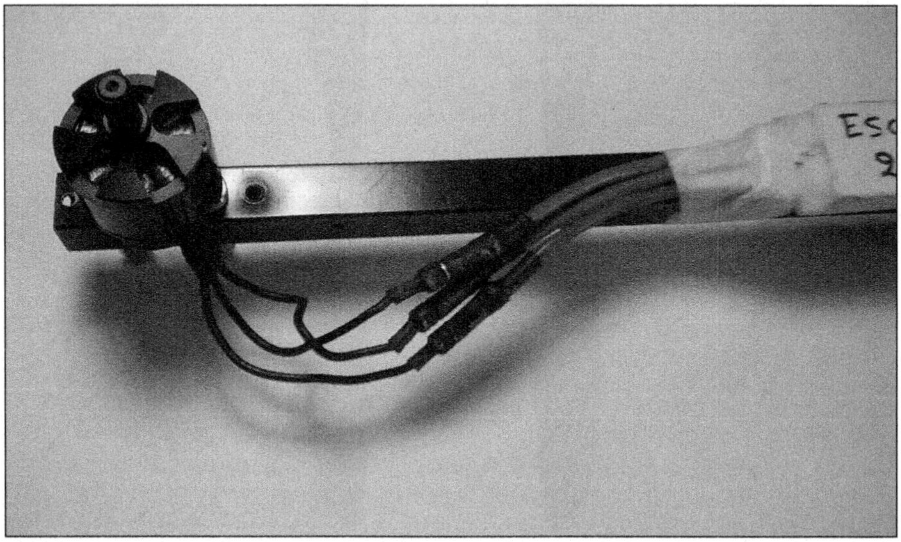

Depending on the flight controller you have, you must connect the four servo wires in the right position. The flight controller that we will use is APM v2.8 and you can find the following figure on the official website of ArduPilot.

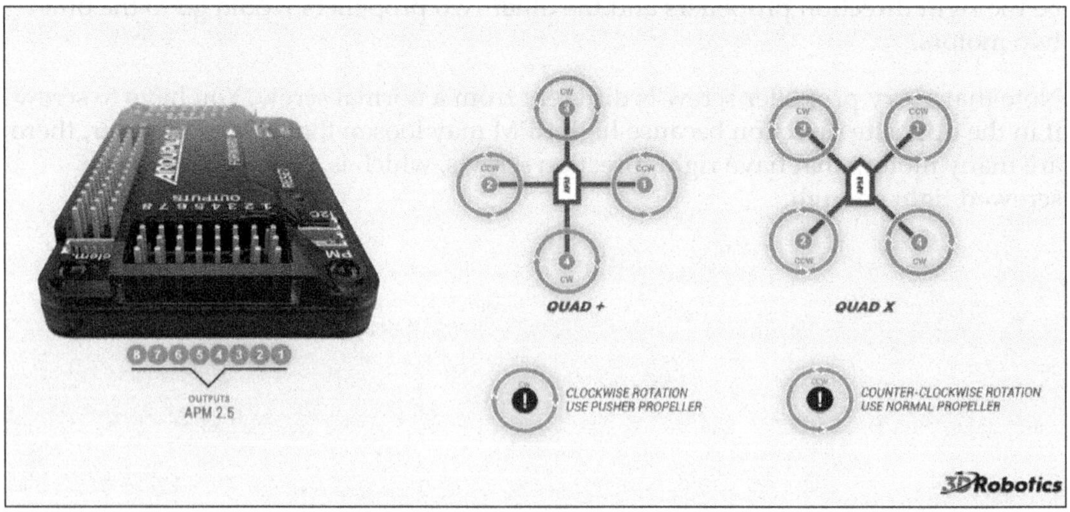

http://www.ardupilot.com

Lastly, the power distribution board, or PDB, must have four female T connectors and one male. Plug in the four female connectors to each ESC and leave the male for the battery. The battery is the last component that should be connected.

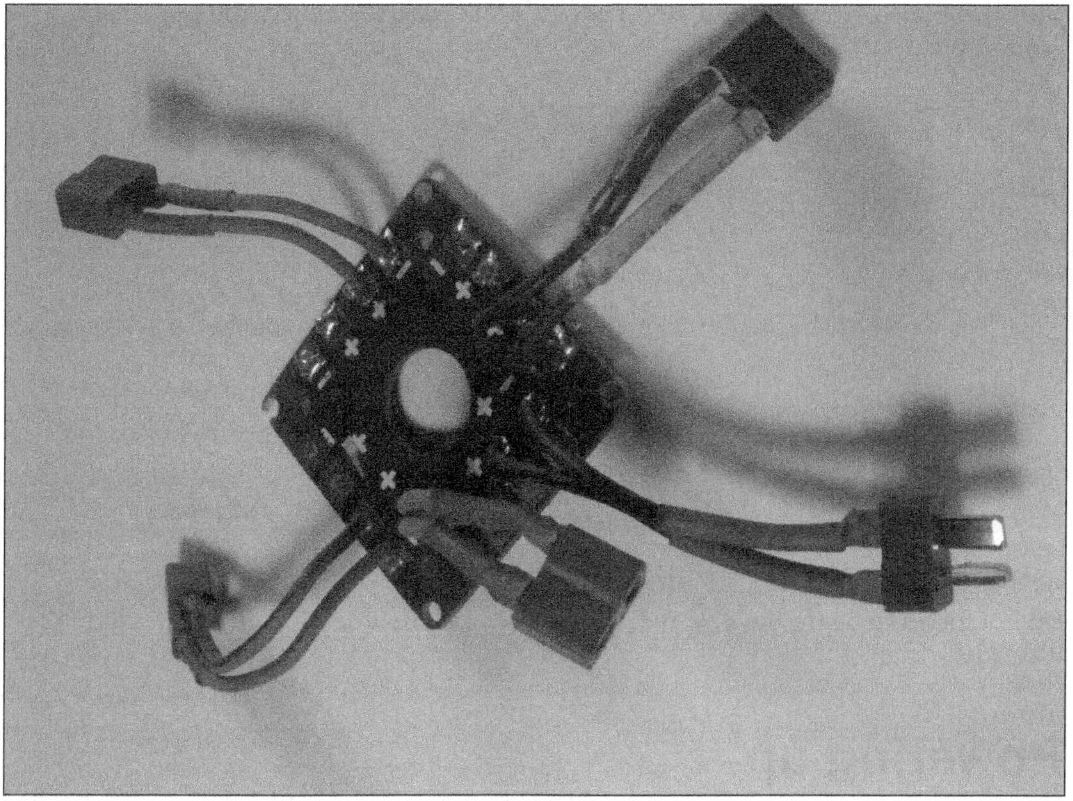

Connecting the receiver

The receiver usually has four to eight channels. Only four of them are necessary for this step. In the following figure, we can find the order in which you must connect your throttle, aileron, elevator, and rudder. Make sure to connect the power and ground pin in the right manner.

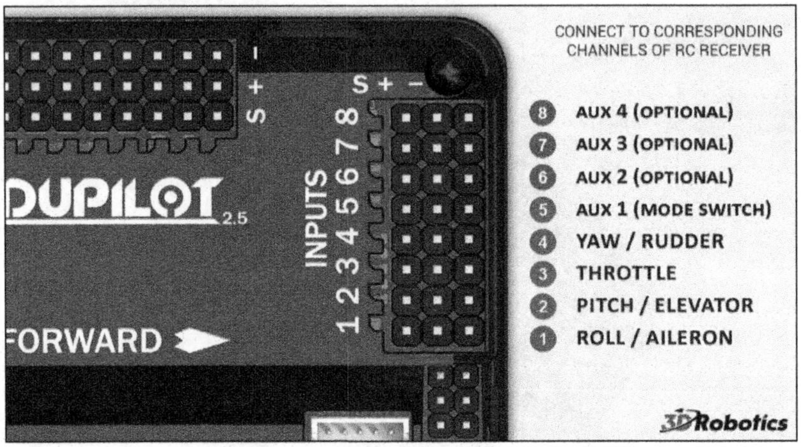

http://www.ardupilot.com

Do not forget to tie up the ESC with electrical tape or something else that will make it steady.

Powering up

Now that everything is connected together, there is only one thing left. Place the battery near the PDB and connect the two T connectors with caution. Note that at this point your quadcopter is not ready to fly at all. You have to pass the calibration and the pre-flight testing list before you can actually fly your vehicle.
The point here is just to make sure that everything is working.

A well-built quadcopter should have beeps from the ESCs and a red-black LED blinking at the flight controller. If you can hear the beeps and see those LEDs, you did a great job!

Summary

All the electronics that we saw must be mounted with caution. Many hobbyists get excited with the idea of flying and make grave mistakes at this step, such as connecting the power to the ground and the ground to the power. As a result, they burn their electronic parts. Be patient and if something doesn't make sense to you, look it up on Google. To conclude, at this point we have mounted the motors on the frame, the propellers on the motors, and connected our ESCs to the motors and the flight controller.

In the next chapter, you will learn how to setup your ArduPilot flight controller and connect everything together before making your first flight.

6

Flight Controller Setup

In the previous chapter, we installed all the electronic parts and, after connecting the battery, we verified that everything was working. The quadcopter needs to *sense* the environment. We need to calibrate our transmitter, accelerometer, and any other sensor that we might have. That way, the quadcopter will be able to *understand* if it is looking to the front, left, back, or right; it can also determine the necessary angles, which will later determine the exact maneuvers required for stabilization. In addition, proper calibration of the flight controller will result in a safe and stable flight. ArduPilot (APM FC) can be connected to a computer with a USB or a MicroUSB cable. This is the right way for all the calibrations that we need to do.

In this chapter, we will cover the following topics:

- Information about the structure of the Mission Planner software
- Uploading software to your ArduPilot flight controller
- Setting up your ArduPilot
- Tuning your flight controller

About Mission Planner

The program that we will be using to set up and calibrate our flight controller is called Mission Planner and you can find and download it from the official website of APM.

Download the Mission Planner software from the official website, as mentioned at `http://ardupilot.com/downloads/`.

After installing the software, you should be able to start the program and you will see the following interface, or something similar to this in case you download a later version of the software.

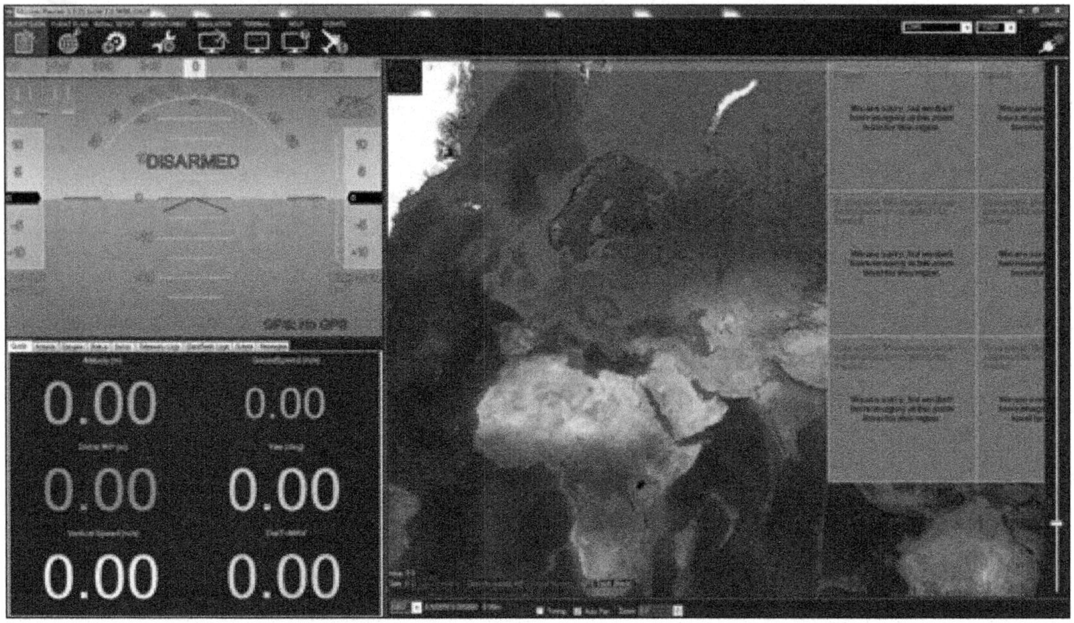

It may look like complicated software but after going through some installation procedures that you will read further in this chapter, it will become quite familiar to you. At first, let's analyze its functionalities and let's have a look at how the software is organized. The main menu is as shown in the following screenshot:

Let's look at an overview of the tabs present in the menu of Mission Planner:

- **FLIGHT DATA**: This tab gives you information about the speed, altitude, horizon, and a map of the location of your quadcopter according to GPS.
- **FLIGHT PLAN**: This tab gives you information about various things when preparing your quadcopter for flight plans (for example, navigation by points on a map).
- **INITIAL SETUP**: In this tab, you can set up your vehicle exactly as you want, go through the calibrations of the transmitter and accelerometer, and so on.

- **CONFIG/TUNING**: This tab consists of parameters and variables that can be changed to affect your stabilization system and other extra functionalities.
- **SIMULATION**: Mission Planner is equipped with simulation software for your real vehicle, where you can simulate anything you want before flying the real one.
- **TERMINAL**: This tab is another way of setting up and calibrating anything you want. It provides unlimited access to almost everything in the Mission Planner software. It is intended for Linux users who are familiar on the Linux terminal.

Uploading firmware

Before connecting the ArduPilot by clicking on the top-right corner, we need to upload the firmware to ArduPilot. The firmware is actually all the blocks and files according to which your quadcopter can fly. The code behind the magic firmware is open source and can easily be found at GitHub (`https://github.com/diydrones/ardupilot`). If you are an advanced programmer, you can join the community and help the ArduPilot flight controller to become better.

You have two options here. Either click on a figure and upload the latest firmware for your quadcopter, octocopter, and so on, or click on **Previous Versions**. After fetching what is necessary, you can upload a previous (maybe more stable) version of the firmware. But first of all, try to upload the ArduCopter V3.2.1 Quad firmware:

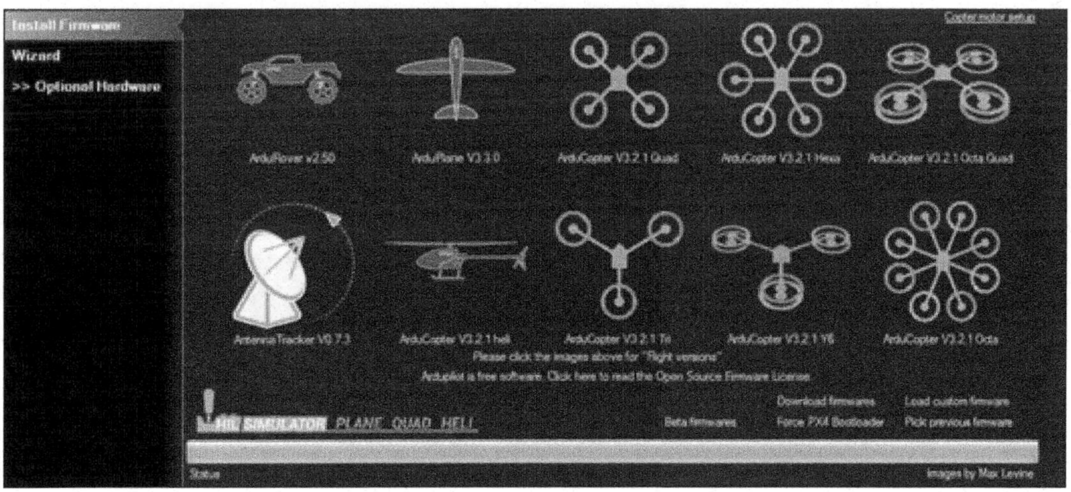

Connecting ArduPilot

ArduPilot can be easily connected to a computer via a USB or a MicroUSB cable. In Windows systems, ArduPilot is usually connected to `COM3` but this actually depends on the devices that are also connected, so it might be a different `COM`. As you can see in the following screenshot, since I have more devices connected, ArduPilot is connected to `COM5`.

For the purpose of troubleshooting, please have a look at the **Device Manager** section that you can find in the **Computer Management** program:

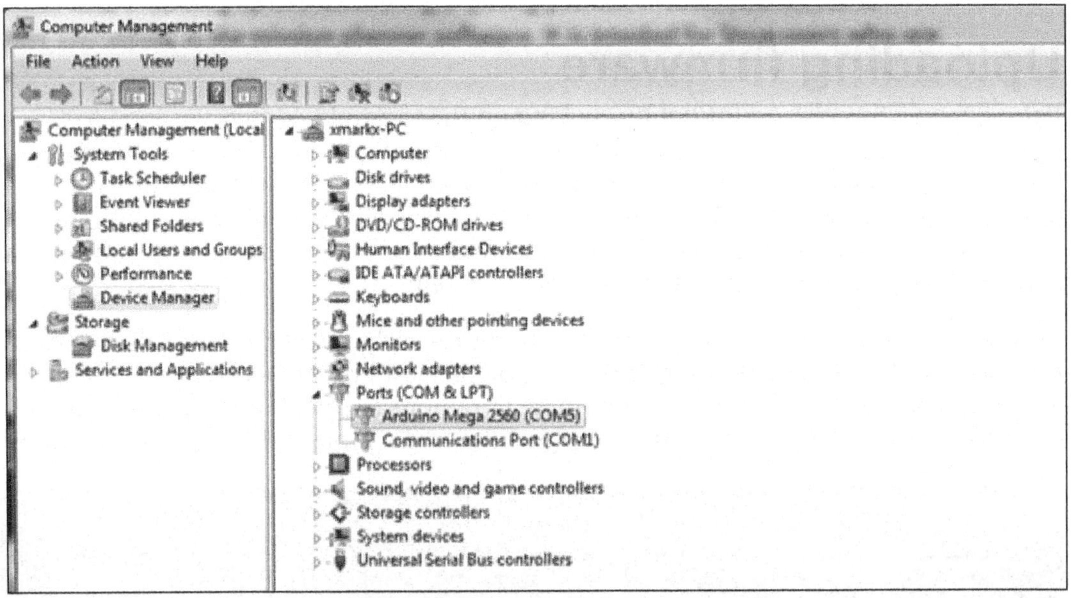

On Mission Planner, you can now see that the selected port is **COM5** and the bound rate is **115200** (default):

As soon as you click on **CONNECT**, there are several things that will happen. First, the software will load all the current parameters that your ArduPilot already has (it might take some time). After that, information in the bottom-left section will appear, as seen in the following screenshot. Also, you can now see the horizon of your quadcopter and maybe some error messages:

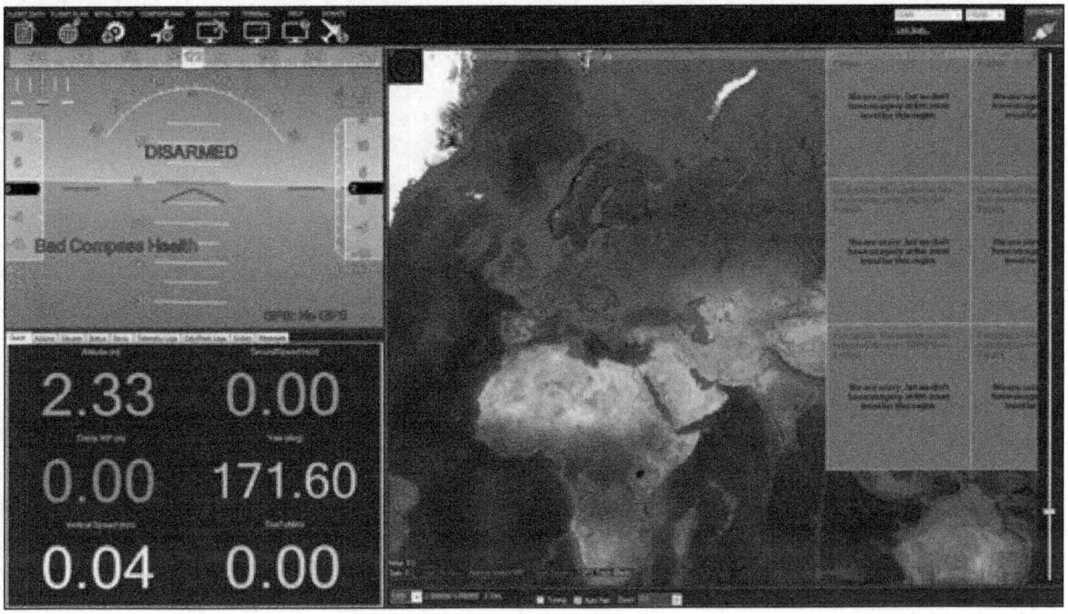

Note that when the figure was captured, my GPS was not installed and an error message appeared over the horizon section. As you can see, there is another message about my compass. I am using ArduPilot v2.8, in which they have removed the inner compass and I have not connected an external one, so to make my quad work, I had to disable the compass checklist. This is something that we will check later.

Calibrating ArduPilot

Now we are ready to calibrate the flight controller. If you click on the **INITIAL SETUP** tab there will be a new list, as shown in the next screenshot:

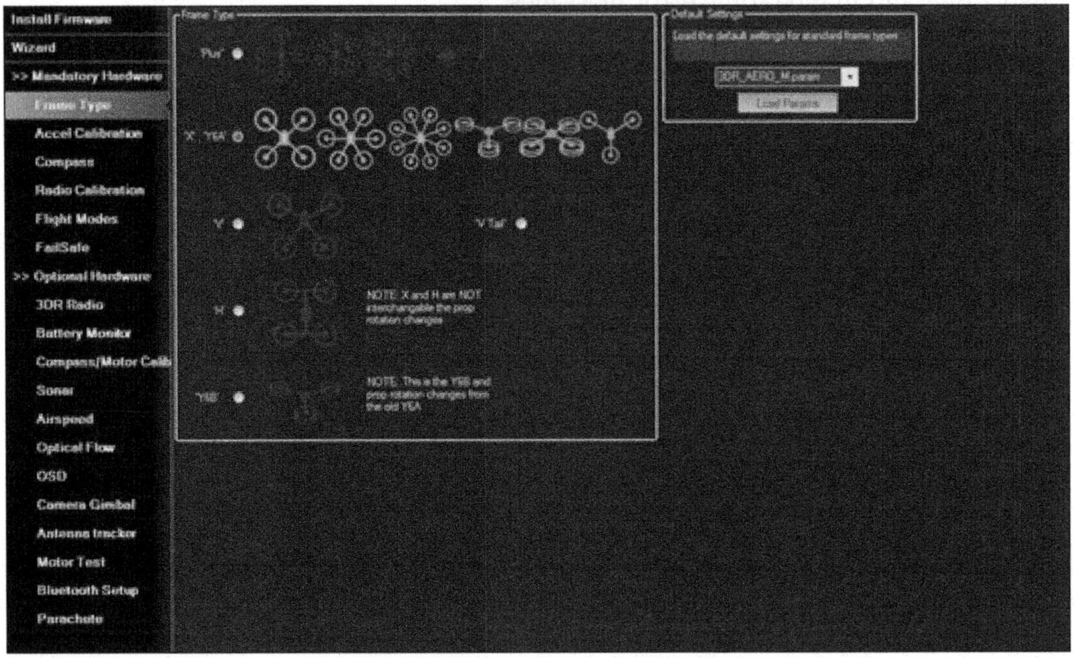

In the **Mandatory Hardware** section, we can calibrate anything we want.

Step 1 – frame type

Click on the **Frame Type** option and choose your frame if not previously selected:

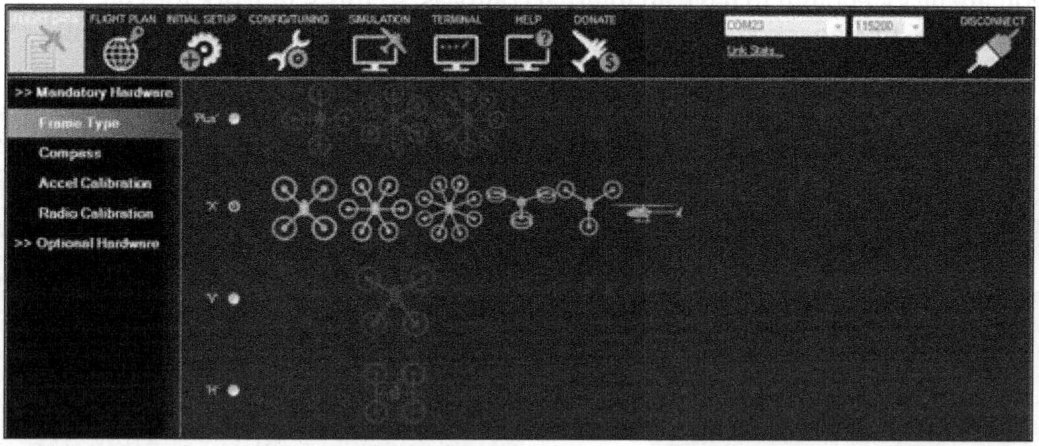

Step 2 – accelerometer calibration

To calibrate your accelerometer, you need to have something with a 90 degree angle because you will need to set this vertically and it's better to set the device as precisely as you can. The acceleration calibration will ask you to place the device as it is; so place the flight controller on a stable table. After that, click on **Next** and place the controller in various different positions according to the instructions in the Mission Planner software. Just make sure to check which side is where in the flight controller.

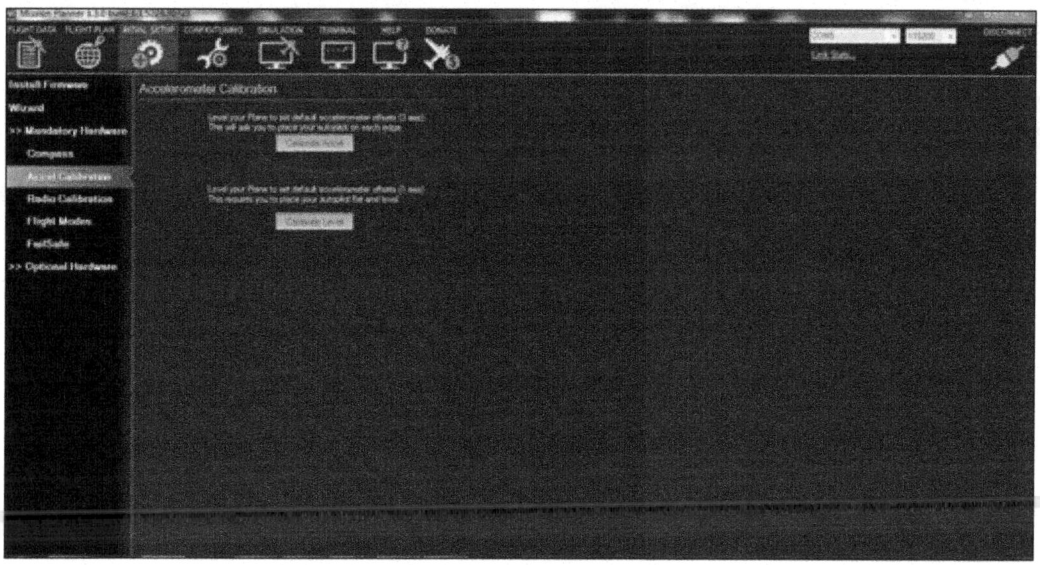

Step 3 – compass calibration

The next step is compass calibration. If you have an earlier version than 2.8, maybe you are lucky enough to have an internal compass, so this will work for you. From version 2.8, the inner compass was removed, so don't do anything since we will disable this feature later.

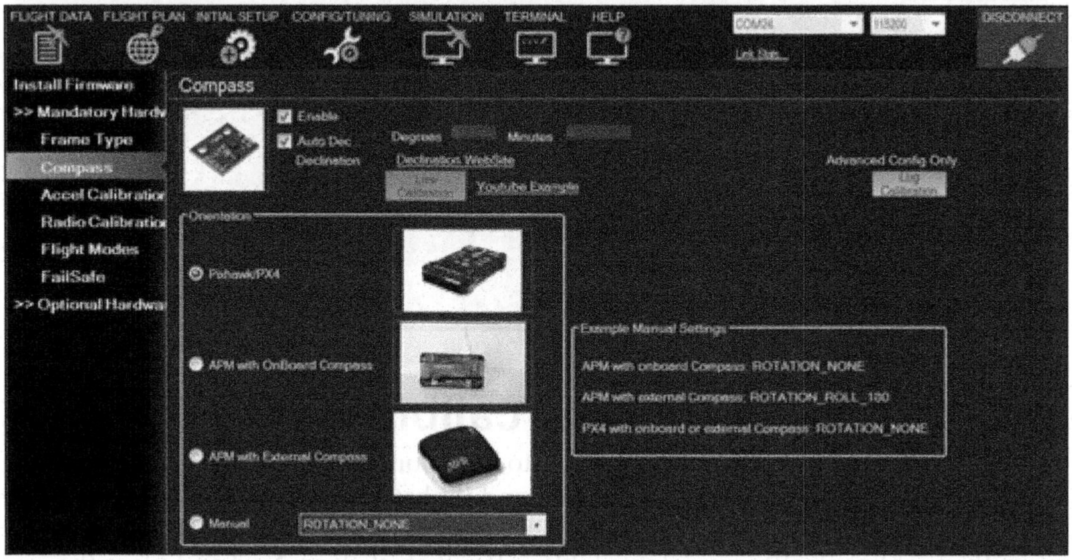

Step 4 – radio calibration

Using the **Radio Calibration** option, your ArduPilot will *learn* the minimum and maximum values of your transmitter. Just click on **Calibrate Radio** and move all your transmission sticks from the minimum to the maximum value. You should see some red lines crossed in the calibration software.

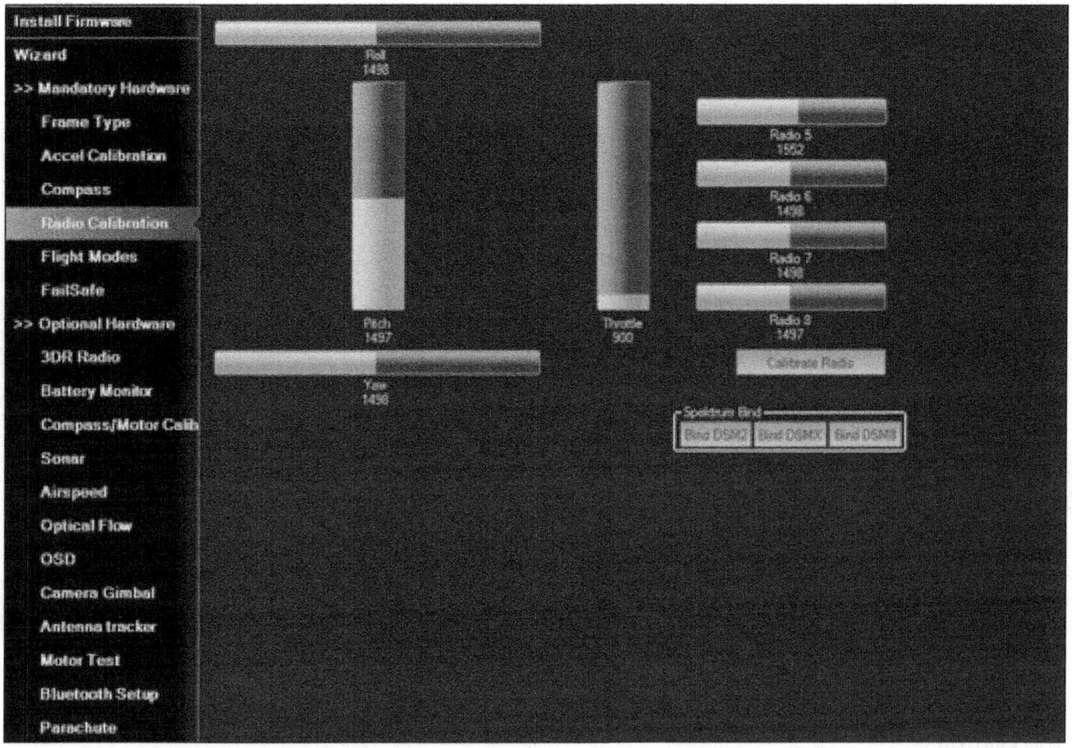

Now, we are almost finished. But there are a few more things that we need to set up. As we said earlier, if your ArduPilot version is 2.8 or 2.6, there will be no compass. This is a problem and your flight controller will never arm itself. As a result, your motors will never spin. This happens because as soon as the flight controller realizes that you want to arm it, it goes through a check list to see if everything is OK before letting you arm the craft and spin the motors. We will need to disable the compass check using Mission Planner, so the flight controller will bypass that check and let you arm itself.

Connect your ArduPilot and go to the **CONFIG/TUNING** tab. Then click on **Full Parameter List** and, on the right side, click on **Find** and search for the word `COMPASS`. You should get the following result:

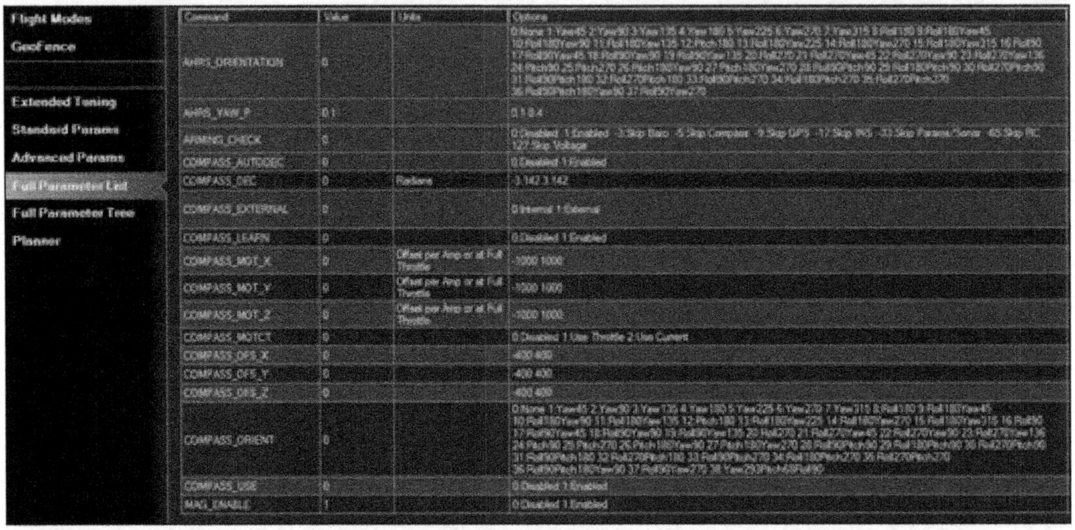

Change the value from 1 to 0 in the **COMPASS_USE** field. This disables the compass.

Summary

In this chapter, we went through the Mission Planner software, which is free, very powerful, and easy to use for ArduPilot flight controllers. Unfortunately, not every flight controller has its own software and is obviously not compatible with Mission Planner. So, in case you have chosen another flight controller, you have to find out another way of doing everything we discussed. It should be clear that anything you buy must be calibrated and configured according to your setup. After going through the setup procedure, you will be able to arm and have a stable quadcopter.

In the following chapter, there will be instructions about your first flight and where and how you can fly safely without risk of damaging your quadcopter.

7
Flight Instructions

In the previous chapter, we went through almost every configuration that was necessary for your first flight. At this point, your flight controller should be ready. In this chapter, you will make some final adjustments and check whether everything is working properly before your first flight. Then, you will find a list of important checks that you should do before you fly your quadcopter.

In this chapter, we will cover the following topics:

- Preflight checklist
- Motor checks
- Securing connections
- Voltage checks
- Weight balance checks
- Location and environment
- Signal interference
- Long distance flights
- Unscheduled landings

Preflight checklist

Even the most experienced pilots have a preflight checklist in the back of their heads. The reason for making and flying a craft like this is for fun or professional business. The last thing anyone wants is to be exposed to situations where things go crazy. So, let's gather some important notes about the things that should be checked before you fly:

- The airframe/landing gear is not damaged
- The props are secure, undamaged, and spin in the correct direction
- The motors are secure and undamaged
- The ESC is secure and undamaged
- The GPS, telemetry, receiver, and transmitters cables are secured
- The battery and other power connections and the voltage are correct

Notice that professional pilots who have bought a DJI Inspire quadcopter or any other expensive craft usually sit down and write a list, which they should always check before flying. It is a good idea to sit down and write a list according to your own quadcopter. The list depends on how many components and what stuff you have added to the craft over time. An example of a checklist for a market drone can be found in the following image:

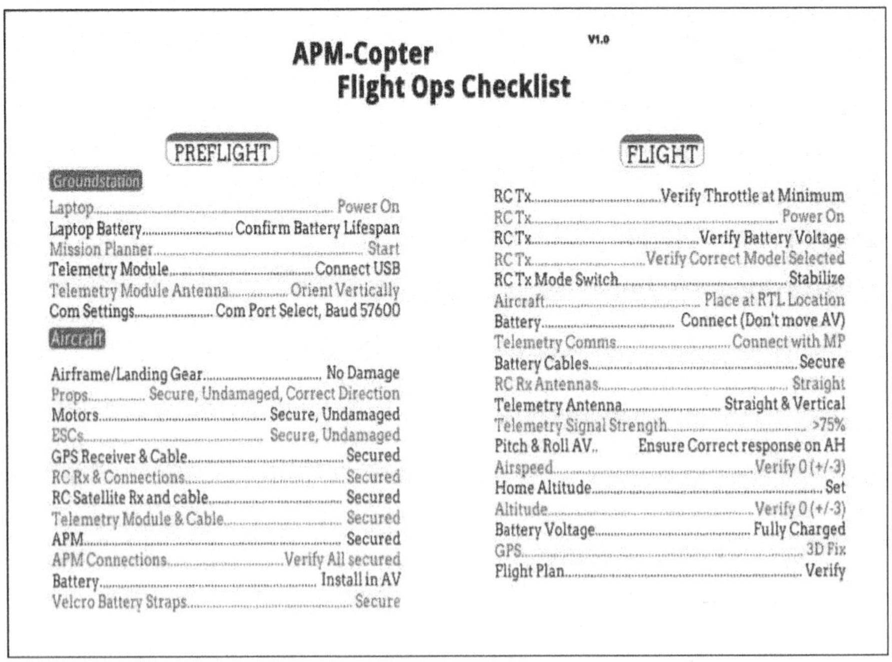

V1.0

APM-Copter Flight Ops Checklist

PREFLIGHT

Groundstation

Laptop........ Power On
Laptop Battery........ Confirm Battery Lifespan
Mission Planner........ Start
Telemetry Module........ Connect USB
Telemetry Module Antenna........ Orient Vertically
Com Settings........ Com Port Select, Baud 57600

Aircraft

Airframe/Landing Gear........ No Damage
Props........ Secure, Undamaged, Correct Direction
Motors........ Secure, Undamaged
ESCs........ Secure, Undamaged
GPS Receiver & Cable........ Secured
RC Rx & Connections........ Secured
RC Satellite Rx and cable........ Secured
Telemetry Module & Cable........ Secured
APM........ Secured
APM Connections........ Verify All secured
Battery........ Install in AV
Velcro Battery Straps........ Secure

FLIGHT

RC Tx........ Verify Throttle at Minimum
RC Tx........ Power On
RC Tx........ Verify Battery Voltage
RC Tx........ Verify Correct Model Selected
RC Tx Mode Switch........ Stabilize
Aircraft........ Place at RTL Location
Battery........ Connect (Don't move AV)
Telemetry Comms........ Connect with MP
Battery Cables........ Secure
RC Rx Antennas........ Straight
Telemetry Antenna........ Straight & Vertical
Telemetry Signal Strength........ >75%
Pitch & Roll AV.. Ensure Correct response on AH
Airspeed........ Verify 0 (+/-3)
Home Altitude........ Set
Altitude........ Verify 0 (+/-3)
Battery Voltage........ Fully Charged
GPS........ 3D Fix
Flight Plan........ Verify

Motor checks

As far as the motors are concerned, we need to check three things. Firstly, we need to check whether all four motors spin. If you check this via the software you are using, there would always be a tab for the motors. Usually, you can spin your motors through Cleanflight, Mission Planner, or any other software. Another simple way of doing this is to unplug your ESC servo wire and connect it to your receiver throttle pin. Since everything will power up as normal, you can increase your throttle and, hopefully, spin the motor that is connected to your ESC.

The second thing that we should check is whether the brushless motor spins clockwise or counterclockwise. We can change it by simply changing the connection between the motor and the ESC. There should be three wires, as shown in the following image:

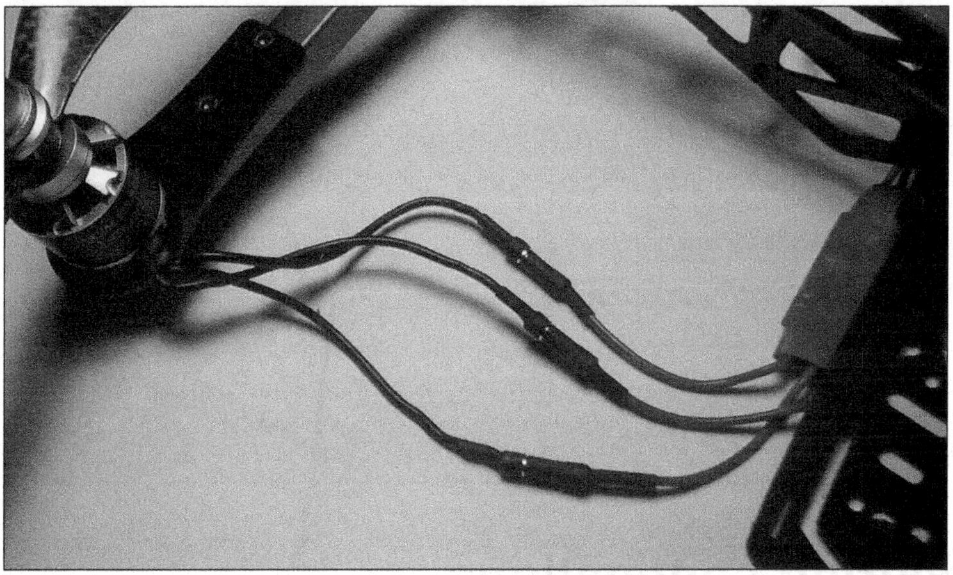

If we disconnect two of the wires and connect them to a hitch, the motor will turn in the opposite direction. Remember that the motor order in the APM flight controller is as follows (CW means clockwise, CCW means counterclockwise):

- Front left motor: CW
- Front right motor: CCW
- Back left motor: CCW
- Back right motor: CW

> Place a small piece of electrical tape in each motor so that you can determine the direction the motor turns. Do *not* test this with your propellers mounted. Try to give a small amount of throttle and watch the direction.

Lastly, the third thing that we should check is whether the propeller is mounted in the right way. A normal propeller has two sides out of which one produces the required thrust. The thrust should be pointed downward. In other words, it should point to the ground. The propeller must be mounted in such a way that the thrust will point in the right way; otherwise, the craft will never hover.

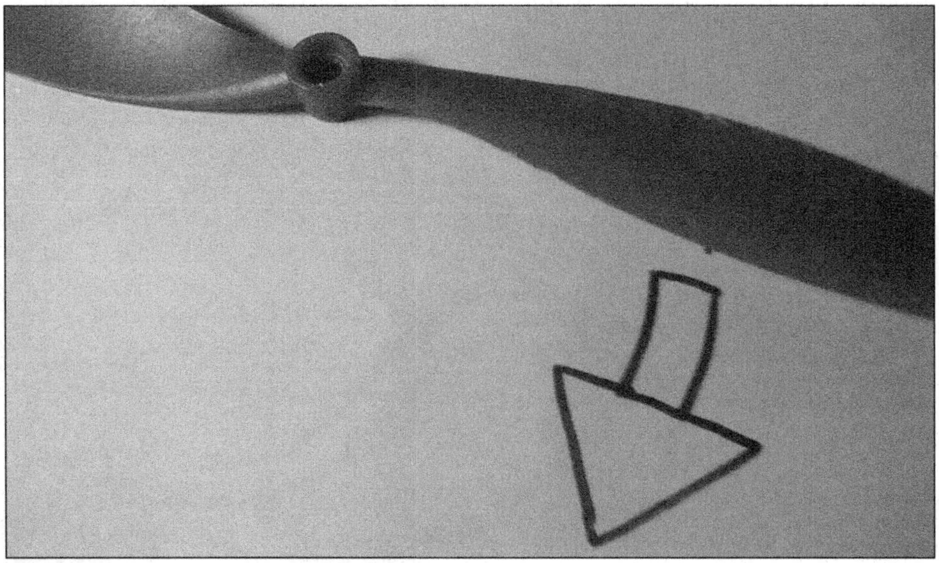

Securing connections

When soldering a connection, it's very important to check whether it is good or it needs more solder wire. If the wires are not well connected to each other, the current will not be able to flow as it should. As a result, your craft will not work properly. Try and check in each connection whether the wires are connected as they were when you soldered them.

If there is any problem, disassemble the craft and solder the wire again. Note that if you don't disassemble the craft, there is a danger of soldering or cutting some other wire. You may just add another problem.

Transmitter voltage and settings

Your transmitter uses a battery. Unfortunately, someday it will have low voltage and it will not be enough to transmit a signal. You obviously don't want that to happen. Check your transmitter voltage as soon as you turn it on. Voltages over 10.5V should be enough for your flight. Some advanced transmitters will beep as soon as they have low voltage, and this means that you have to land your craft as fast as you can.

When the voltage is lower than 10V, you should land your craft as soon as you can and power up your transmitter with another battery.

Balancing the weight

Market quadcopters have specific positions for the battery and every single component that you can add later. DIY-built crafts, however, need to have their weight centered. This is crucial for a stable hover. If there is more weight at the front than the back side, once you hover, you will clearly see that there is a small drag on the front side. Yes, it is possible to eliminate this drag using the trim in the transmitter. But why change the settings when you can simply center the weight before you start flying? Let's see how we can do this.

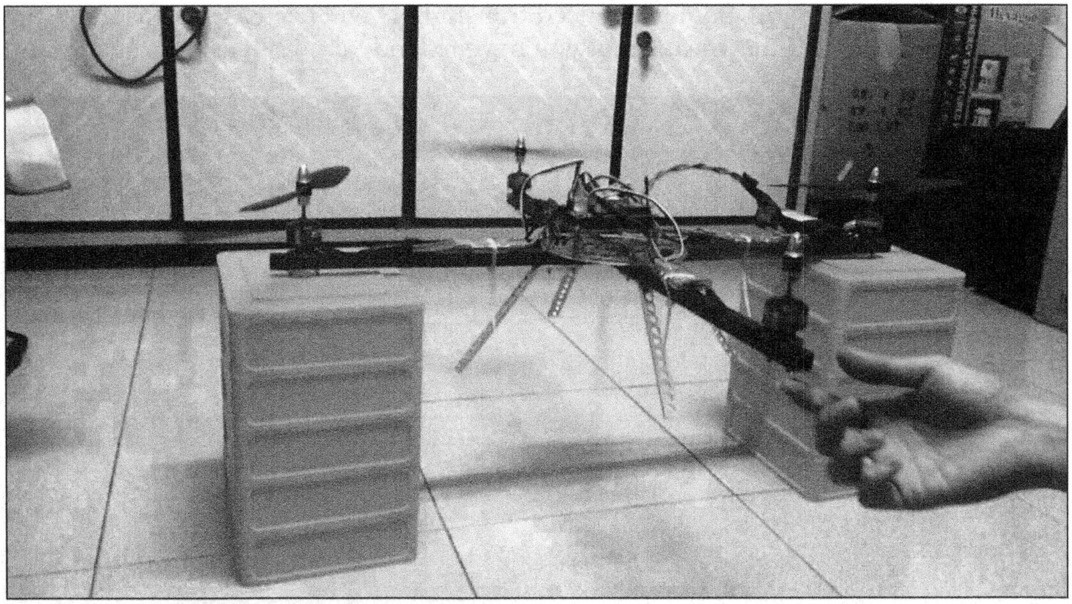

You can try and do something as shown in the preceding image or lift your quadcopter using your two hands. Gently grab the edge of one arm and, with your other hand, gently grab the edge of the opposite arm. This way, you will be able to eliminate any drag along the axis from your left to your right hand. Your quadcopter should be stable. If you see one of the other two arms falling or being lifted, you have to change the position of the battery or anything else that you have mounted. Repeat this process until the weight is balanced. When it is stable, gently grab the other two arms in the same way and, finally, you should end up with a stable quadcopter with almost minimum drag.

Location and environment

If you have never flown a copter before, you may not understand the importance of the flight location. A good location provides you with the following:

- A better view of your vehicle
- Accurate and stable control
- Less signal interference
- More emergency plans

If you think about it, all of the preceding aspects are crucial. The more bullet points that are missing, the more possibility you have of ending up with a crash.

In the following image, you can see that there are no people and objects that can harm or make our flight difficult. This is an excellent location to test your quadcopter and to also learn how to fly:

http://www.rcmodels.com/

On the other hand, crowded places, as shown in the following image, will make your flight difficult and insecure and increase the possibilities of a crash or injury:

http://www.pps.org/

Do not fly over crowded places. Even if your craft costs 1,000 dollars and you are a pro, a craft is a machine. It consists of hardware. As your computer can someday crash, your craft can crash too and do something that you don't want it to. If you lose control, you will have a serious problem.

Usually, pilots tend to add a parachute to their craft so that it can drop slowly.

Keep in mind that the location and the environment in which you are going to fly your craft are crucial factors for a successful flight. Always try to avoid crowded places with kids playing around. Your craft should never hover above people at low altitudes even if you are the best pilot in the world. Accidents have occurred and the injures were serious.

Also, if the weather is cloudy and windy, your craft will have a small drag in the direction of the wind. This can be countered using the trims in your transmitter. However, if you change the yaw of your craft, your trims will make it worse. To conclude, try and choose days without clouds or wind. This way, you will have a pleasant and smooth flight. Lastly, never fly with your eyes toward the sun. This will make you lose sight of the craft and, unfortunately, you will end up crashing.

Signal interference

Your transmitter communicates with your receiver using signals. A signal is an electromagnetic wave. As you can understand, everything that uses electromagnetic waves is a possible interference to our communication, which is why we call everything else apart from our data as noise. TX/RX communication has some ways of eliminating noise. However, if the signal has a high frequency, we have a problem. You may still be able to hover and control your craft, but if you are using goggles or have any kind of FPV in an external monitor, you will not be able to view anything at all.

Thankfully, there are two free bands where antennas transmit data through 2.4 GHz and 5 GHz. Everything that is working in 5 GHz does not affect your transmitter as almost all TX/RX use 2.4 GHz bands.

Keep in mind that signals are waves in the air; if they hit a building, they reflect and come back. A signal can become very weak if it needs to pass through a building. This means that the receiver will receive a low signal and your failsafe mechanism might get activated. The preceding scenario is not a crazy one. For example, if you want to explore a building using a small quadcopter, your receiver will not get the full signal when your craft is inside the building.

Long distance flights

Some people tend to make long range flights with a simple transmitter-receiver setup and nothing more. Be careful if you also try this because not only is this extremely dangerous and increases the risk of a crash, but you may lose your craft forever. As we mentioned earlier, signal interference is a huge problem when you have a long flight because you need to know what is in front of you and above you. If the area is where you live and you already know the buildings and other structures, then you obviously have no problem. But otherwise, keep in mind that you may lose vision at any moment.

There are many unexpected dangers that even a pro pilot cannot predict. If you want to do this for fun, do not do it. If, for any reason, you want to make it happen, then be very careful, especially if you are flying through FPV. You will have so much noise that your monitor will be useless for a few minutes.

Unscheduled landings

Let's say that your craft is somewhere in the air. Something is happening and you are losing control. Maybe your battery has low voltage or some of the screws have become loose. Actually, it doesn't matter what exactly is happening over there. You have to act and act fast. If your craft is two or three meters over the ground, in most cases, you minimize the throttle and land within 2-3 seconds. However, if the craft is at 10 or 20 meters of altitude, there are various factors that you should be thinking of. Here are some thoughts that a professional pilot could have:

- How much time do I have?
- What's the nearest ground that I could land on?
- What if I land it over a tree?
- Are there any people nearby?
- What failsafe mechanisms do I have?

These are some of the questions that you should answer in 2 seconds. Landing on a tree or some other crazy location can be avoided. However, there is a rule, which is quite important. If you are going to crash or if you have lost control, you must do whatever you can to end up with the least damage. Sometimes, you cannot avoid the damage, so you should always try to minimize damage even if this requires you to break a motor or something else.

Summary

As you saw, there could be several unexpected issues while flying your craft and you have to be extremely careful every time. If anything goes wrong, do not keep flying and land as soon as you can for your safety and the safety of your craft, anyone else, or anything else nearby. In the next chapter, we will talk about cameras and first person view. We will also talk about servo mechanisms and other extra components that you can easily mount on your drone.

8
Cameras and Extra Functions

You should take time to learn how to fly and land your quadcopter. As soon as you add a high cost camera or any other equipment, you have to be able to avoid crashing again. Apart from the camera, there are a lot of extra things that you can add to your craft. Each of them should have a specific reason to exist but it is obvious that your craft's functionality is limitless. In this chapter, we will go through some basic components that you can add to your craft to do some more advanced tasks. Each of the following extra components can be used to capture photos or add a failsafe mechanism. Understand that almost everything in the following list has some weight. Weight is the worst nightmare for an aerial craft. So, if you are going to use any of them, be sure that you really need . The worst case scenario is to have a quadcopter whose weight is more than the thrust that all your motors can provide. There is no way to lift this craft. But even if you lift it, you don't want to hover with 80% of the maximum thrust. Your craft will be unstable and your flight time will be from 6 to 7 minutes.

- Camera for photographs
- Camera for FPV (first person view)
- Failsafe parachute landing
- One or more servo mechanisms
- LED strips
- Arduino boards
- GPS system

Camera for photographs

Whether it is a simple webcam or a professional GoPro camera, you can easily mount it on your quadcopter and capture videos or photographs of your flight. There is no one-step guide for how to place your camera on your quadcopter. What you will read next are some basic issues while placing your camera and some things that you should always avoid.

http://phys.org/

Position in relation to other components

Assuming that your build is an X shaped quadcopter, your camera should be placed at a position near the center of mass.

Usually, most cameras weigh more than 100 grams and this weight is enough to change the craft's center of gravity. As a result, depending on the position of the camera, your craft will have an imbalance in weight causing decreased performance and battery life. This is something we can avoid by keeping the camera position as close to the center of gravity as possible. Take a look at a DJI Phantom drone and its camera position in the following image:

http://www.dji.com

Another issue that we should consider is power interference. Every wire starting from the battery to each component has interference and if the camera is near the 12V wires, it will result in some noise. So you must keep in mind that wires should be at least 1cm away from the flight controller and the camera.

Vibrations

Motors produce more vibration than you think. The vibrations are a source of noise and affect the camera and the flight controller. As far as the camera is concerned, the video that you have captured will be the worst video ever. The flight controller, on the other hand, must have minimum possible vibrations to minimize the error that is produced and to keep the quadcopter stable. There are anti-vibration systems on the market that do a very good job. Here is one of them:

http://www.aliexpress.com

YouTube or other video sharing websites have stabilization algorithms that can somehow produce a more stable video but it's still not the result we want. To minimize the vibrations, we can mount some kind of vibration absorber between the camera and the craft, similar to the ones shown in the following image:

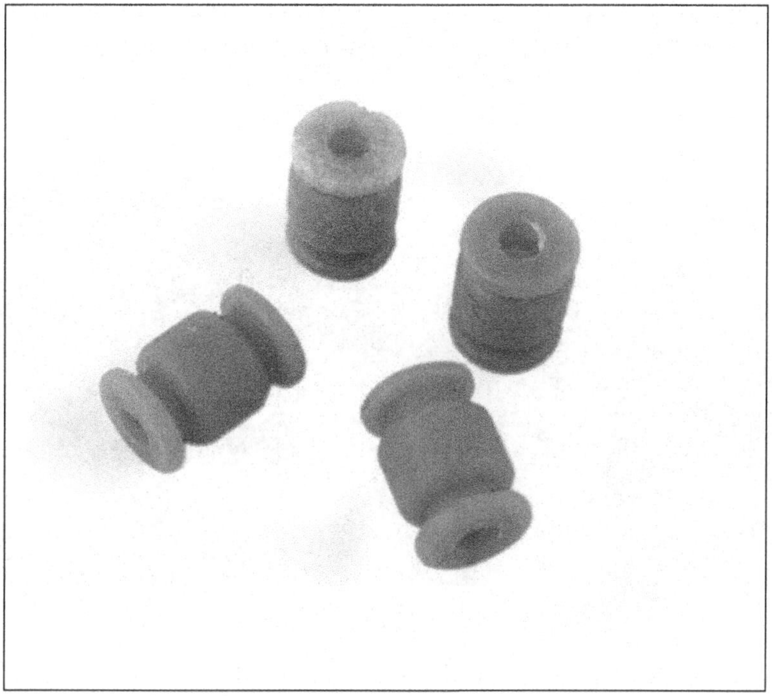

http://i01.i.aliimg.com/

Camera for FPV

First person view, or FPV, is a way of flying your craft that allows you to have eyes in the sky. The *feeling,* and of course the view, is a lot different than flying without FPV. Even if at first it is difficult to control the quadcopter, after a while it is just amazing.

You will need two things to set this up. First, you will need a small camera to mount on your craft. The camera must be connected to an antenna and the antenna will transmit the signal to a ground station antenna. The ground station antenna is always connected to either goggles or a monitor.

Goggles are special glasses that you can wear that provide a FPV in front of your eyes, as you can see in the following image from a YouTube video:

https://i.ytimg.com

This image is what the drone sees and this is also what you see on your goggles. The goggles look similar to the ones in the following image and they are pretty easy to install:

http://www.hobbyking.com

On the other hand, there are many monitors that can be used for such a cause, such as simple FPV monitors. In the following image, we can see a classic FPV monitor:

`http://www.hobby-wing.com`

Most pilots do not prefer very large monitors because you may need to travel to a park or somewhere else and you need to be able to hold your craft, transmitter case, FPV case, and some more stuff. There are many things that you should have and it's kind of difficult to walk with everything in your arms.

Failsafe parachute landing

Landing is usually the worst part of the flight. It's the most difficult one too! The scenario that we are describing here is when the craft has a problem and it is difficult to land or there isn't much time to land safely on your own.

To avoid crash landings, quadcopters can be equipped with parachutes. The APM flight controller has an option of adding a parachute. The hardware mechanism of a parachute must be connected via servo wire to the flight controller and the trigger must be configured through the software.

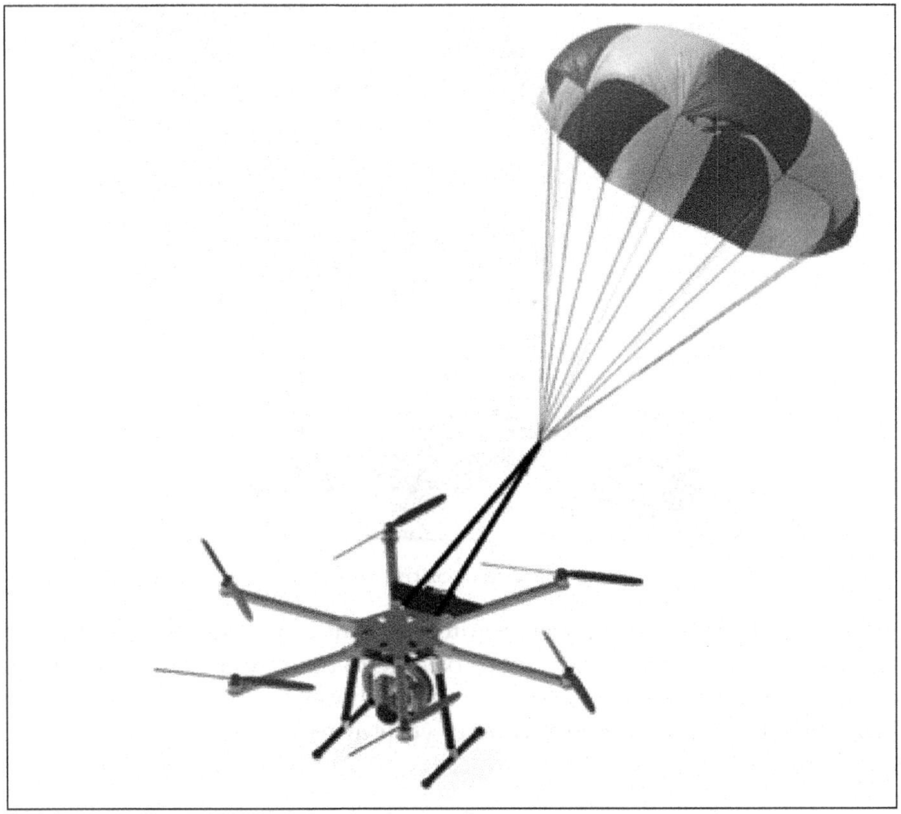

http://www.diydrones.com

Parachutes have a very simple way of activating. The bottom side of the parachute is open and the idea is that while the quadcopter is flying, the air goes from the bottom side and *waits*. As soon as we trigger the parachute, the top side will open and the parachute will fill with air, slowing the craft enough for a safe landing.

Servo mechanism

A servo is a mechanical system that rotates from 0 to 180 degrees. To help you understand the power of the servo mechanism, let's have a look at a common servo and where and how can we use it.

http://www.conrad.com

The preceding image is of a simple servo. Usually, a servo consists of two or three different plastic modules that we can easily change according to our goal. The servo is connected with 5V to any device like an Arduino board or most of the flight controllers that support a servo.

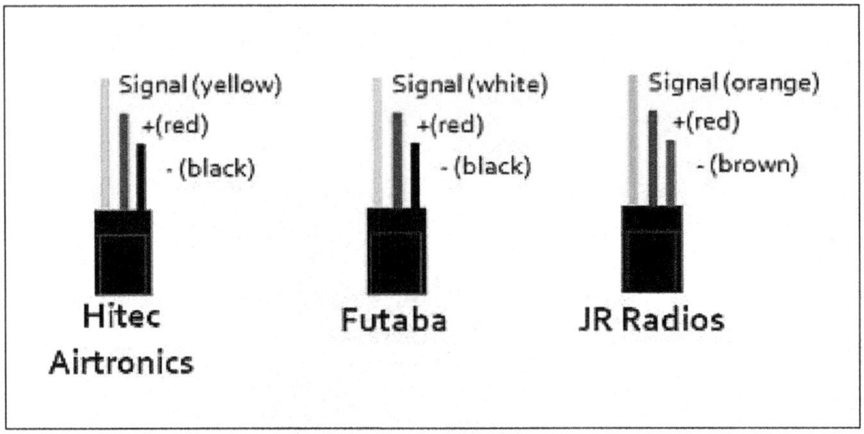

`http://www.imagesco.com`

You have to identify the servo you bought and connect the wires in the right order according to the preceding image.

One or two servos can be used to create a simple DIY grabber at the bottom of the craft. A servo goes from 0 to 180 degrees, so it's quite simple to create a mechanism similar to this. Furthermore, we can use it to trigger various dropping mechanisms.

LED strips

Flying in daylight doesn't require any extra components. The craft is visible hovering from the ground and your FPV monitor (if used) should be very clear as far as the light is concerned. On the other hand, when the sun goes down or the sunlight is low, you may consider adding an LED strip. LED strips are extremely low cost and lightweight. They are a great way of indentifying your craft in a low light environment! In most cases they look like in the following image:

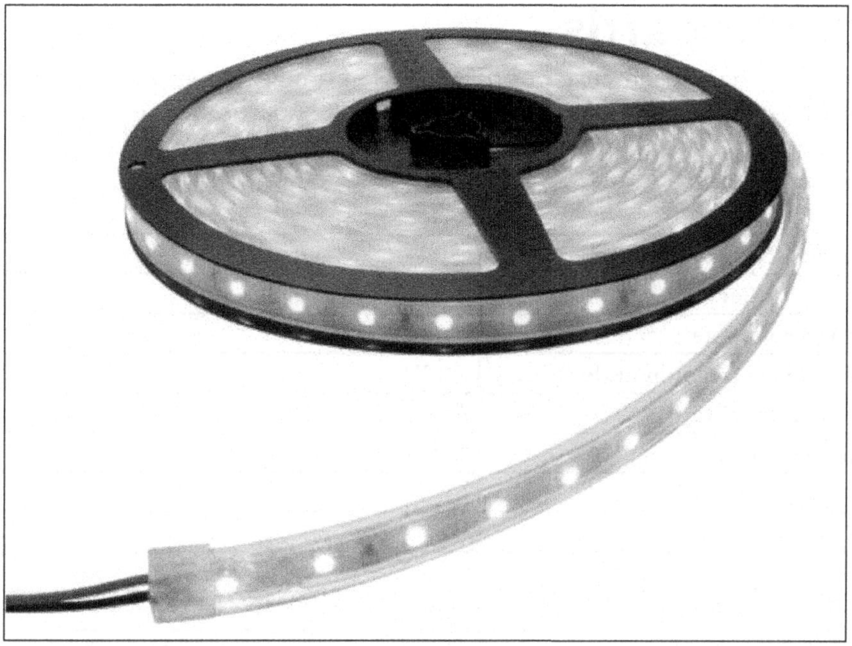

http://www.quickmarinelighting.com/

LED strips can be cut in small pieces of 5-10cm. The input voltage is about 12V, exactly the voltage of the usual LiPo batteries. Adding an LED strip to your craft will result in something like the following image. Obviously, you can understand its importance in a low light environment.

http://www.makehardware.com/2015/07/

Arduino boards

Arduino boards have been very popular. APM and many other flight controllers all run on an Arduino MEGA or other Arduino-like boards. These boards are quite cheap, easy to install and use. The specifications of an Arduino UNO board are as follows:

Arduino UNO specifications	
Microcontroller	ATmega328P
Operating voltage	5V
Input voltage (recommended)	7-12V
Input voltage (limit)	6-20V
Digital I/O pins	14 (of which 6 provide PWM output)
PWM digital I/O pins	6
Analog input pins	6
DC current per I/O pin	20 mA
DC current for 3.3V pin	50 mA
Flash memory	32 KB (ATmega328P) of which 0.5 KB is used by the bootloader
SRAM	2 KB (ATmega328P)
EEPROM	1 KB (ATmega328P)
Clock speed	16 MHz
Length	68.6 mm
Width	53.4 mm
Weight	25 g

An Arduino UNO board looks like this:

`http://www.solarbotics.com`

The specifications of an Arduino MEGA board are as follows:

Arduino MEGA Specifications	
Microcontroller	ATmega1280
Operating voltage	5V
Input voltage (recommended)	7-12V
Input voltage (limits)	6-20V
Digital I/O pins	54 (of which 15 provide PWM output)
Analog input pins	16
DC current per I/O pin	40 mA
DC current for 3.3V pin	50 mA
Flash memory	128 KB of which 4 KB is used by the bootloader

Arduino MEGA Specifications	
SRAM	8 KB
EEPROM	4 KB
Clock speed	16 MHz

An Arduino MEGA board looks like this:

`http://www.solarbotics.com`

Usually, similar to APM flight controllers, if the purpose is to build a flight controller, then Arduino MEGA is the best choice. That's why APM FC is an Arduino MEGA with many other sensors. Arduino UNO can be used as an extra controller or a board that triggers anything using a time clock, a signal, and so on.

For example, an Arduino UNO can be used to trigger a servo mechanism after two minutes of flight. Since Arduino uses a 5V input, any signal from the transmitter-receiver communication can be easily read and used. There are also various shields for Arduino boards out there. A shield is a chip that can be mounted on the top of the Arduino boards, or can be connected with the appropriate pins, which give some input or output to the user. There are shields that can be used to mount a SIM card, wireless transmitters, temperature, humidity, or distance sensors, and much more. For more information, visit Arduino's official website at `https://www.arduino.cc/`.

GPS systems

Being able to know your map location at every single moment is very important for long flights. There are many GPS systems in the market that help your controller and feed it your craft's coordinates. So, you can get your location as a mark on a map via telemetry and use the Mission Planner software or any other navigation software to place some navigation points and set your quadcopter to autopilot.

In the following image, we can see the NEO 6m GPS module. It can be easily mounted on the ArduPilot flight controller and it will feed your craft the required coordinates:

`http://www.aliexpress.com`

Note that the GPS also has drawbacks. Unfortunately, the GPS as a standalone system cannot get accurate coordinates. As a result, the flight controller uses a GPS module combined with an accelerometer, magnetometer, and maybe some more components. Each of them has small errors but the combination is very good. A GPS module should not be trusted under cloudy weather, in the rain, or inside buildings. All these situations lower the strength of the signal from the satellite and, as a result, increase errors.

Summary

As you have seen, there are several other components that you can mount on your craft to increase its functionality. Adding extra components have their drawbacks and they are more weight to lift. So always choose wisely and fly safe. These were just a few of the many exciting things you can add to your quadcopter.

In the next chapter we will see how to maintain and repair our craft in the case of a crash.

9
Repair and Maintenance

Almost every necessary note and instruction for your flight has been mentioned in the previous chapters. However, even professional pilots sometimes crash planes. So, in this last chapter, you will learn how to maintain your quadcopter and, in case of a crash, how to estimate the damage caused and repair any part that is not working properly. Based on the skills learned in the previous chapters, you will be able to completely fix any problem caused by a crash.

The first part of the chapter will be about how to repair a damaged craft and, later on, we will learn some tips and instructions for how to maintain your craft.

In this chapter, we will cover the following topics:

- Crash analysis
- Common issues
- Damaged propellers
- Damaged frames
- Vibrations
- Motor maintenance
- Voltage testing

Crash analysis

Firstly, we have to analyze the crash, sit down, and find out what happened. We need to find out if it was the pilot's mistake, hardware error, or maybe it was because of something in the environment, a distraction or something like that? This sounds silly, but learning about what exactly caused the crash will help you fly safer and with more caution the next time. After all, it's all about knowledge and getting better and better every day by learning something new.

It's very difficult to create a troubleshooting guide about crashes since there are several factors that change many things. For example, if the craft does not act as you want when using your transmitter, the problem is usually low voltage or some bug in the controller, or it may be a hardware problem such as loose screws.

Common issues

Anything can happen, but there are some common problems, and we will first try to focus on them.

Damaged propellers

The damage done to the propellers depends on the actual size of the propellers. If the propeller is flexible, then it usually breaks into two pieces in the middle. On the other hand, if it's not too flexible, then it may break into more than two pieces. As you can see in the following image, the propeller has touched the ground during the flight:

Here, we have two options. Either continue and fly if the damage is not too great or replace the propeller. If you have another propeller, I suggest that you replace it as soon as you can.

Damaged frame

While landing, your frame may have problems, depending on the material and construction. It may lose the weight balance, and you need to fix this in any way that you think will work. It's a good tactic to always check the weight balance before a flight since this can result in a constant drag.

Vibrations

We have discussed some things about vibrations, and you should be able to understand how important it is to eliminate any kind of vibrations in your craft. A good advice is to add something, such as a vibration absorber, as shown in the following image:

It can be added to the flight controller or anything else that will not vibrate.

Maintaining your craft

After each flight, it is important to keep in mind that your craft consists of electronics. If you don't clean, maintain, and keep your electronics safe, future problems will occur for sure. There are some small things that you can do after a flight to have a stable and well-maintained craft.

Firstly, fly and land safe. This indicates that you are a good pilot, and you can land your quadcopter with caution without damaging anything.

Secondly, do not let your battery go below 15% of its capacity. Flight controllers can alert your craft to land as soon as the battery gets too low. This is a good way of ensuring you always have a sufficiently charged battery. If you cannot charge your battery, try and change the program to something such as NiMH for about 2 minutes. Do not charge a LiPo battery as NiHM for over 5 minutes because you will damage it. The goal here is to charge and increase the voltage so that you can bypass any voltage errors and charge it as a LiPo.

Cleaning the motors

Every brushless motor can be cleaned by opening the clip. Gently try to split the motor into two parts by removing its top. This requires some effort as the inside of the motor is a positive and negative magnet. After removing the inside of the motor, you should have something that looks like the following image:

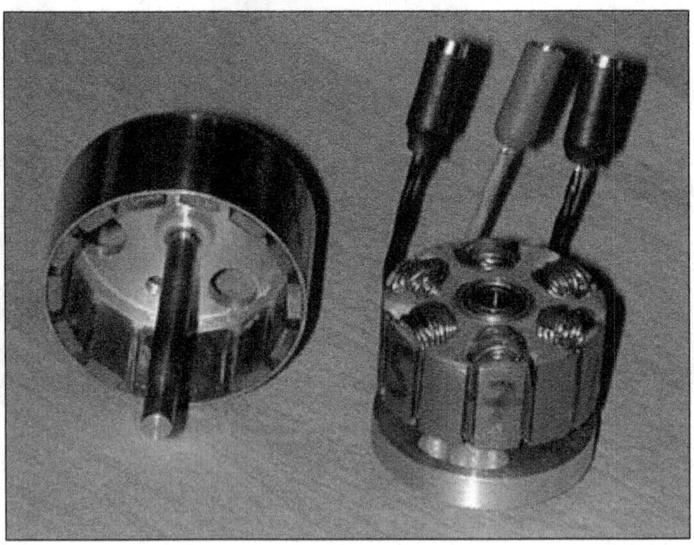

You can clean your motors using a soft doily. You can also use some oil to make the motors spin better.

Voltage testing

Measuring the voltages of your batteries or any other connection is very important to debug issues. You can find out whether a connection has power or not, whether the connection is good, or the voltage is lower than it should be. A multimeter is a tool used to measure the voltage, ampere, and resistance:

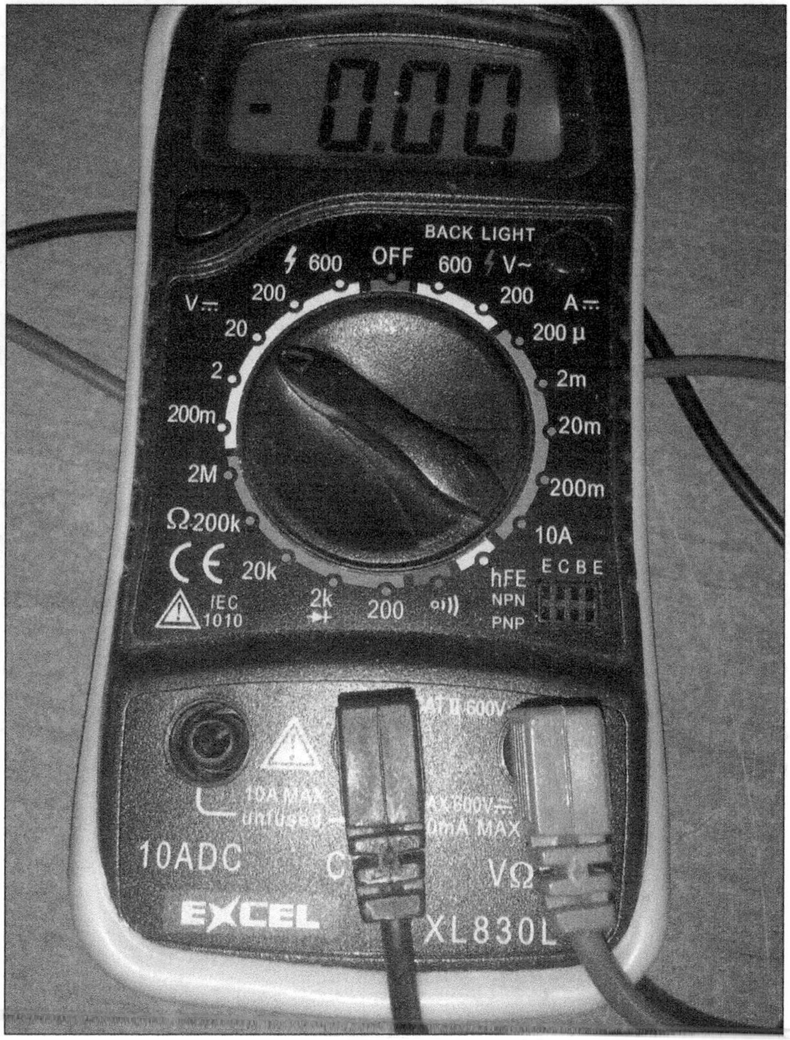

A multimeter has one red and one black wire. The red wire must always be connected to the positive wire and the black wire to the ground or negative wire. For example, you can measure your battery's remaining voltage by connecting the appropriate wires, as shown in the following image:

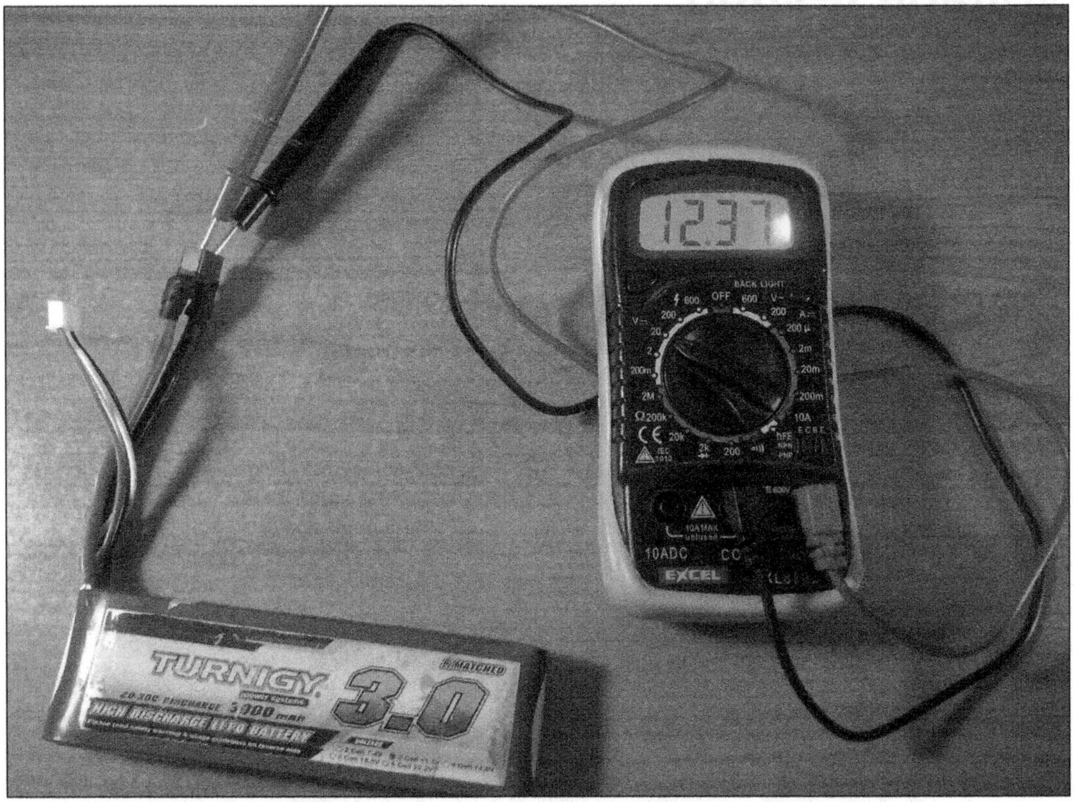

You can use a multimeter to measure the voltage of any connection in your build. For example, if your LED strip does not light up, you have to check whether the connection that goes into the LED has power or not, as shown in the following image:

A simple and easy way is to find the two wires (black and red) that go to the LEDs and measure the voltage.

You will then be able to know whether the issue is with the LED strip or with the PDB. In other words, this tool is a great way to eliminate the possible issues and find the problem.

Summary

This was the last chapter. Hopefully, it is now clear to you that maintaining your quadcopter and learning how to repair any damage is very important. Take time to fix any issues and make it as stable as you can. You will definitely enjoy your flight more.

Index

S

T

U

V

W

[PACKT] open source
community experience distilled
PUBLISHING

Thank you for buying
Building a Quadcopter with Arduino

About Packt Publishing

Packt, pronounced 'packed', published its first book, *Mastering phpMyAdmin for Effective MySQL Management*, in April 2004, and subsequently continued to specialize in publishing highly focused books on specific technologies and solutions.

Our books and publications share the experiences of your fellow IT professionals in adapting and customizing today's systems, applications, and frameworks. Our solution-based books give you the knowledge and power to customize the software and technologies you're using to get the job done. Packt books are more specific and less general than the IT books you have seen in the past. Our unique business model allows us to bring you more focused information, giving you more of what you need to know, and less of what you don't.

Packt is a modern yet unique publishing company that focuses on producing quality, cutting-edge books for communities of developers, administrators, and newbies alike. For more information, please visit our website at `www.packtpub.com`.

About Packt Open Source

In 2010, Packt launched two new brands, Packt Open Source and Packt Enterprise, in order to continue its focus on specialization. This book is part of the Packt Open Source brand, home to books published on software built around open source licenses, and offering information to anybody from advanced developers to budding web designers. The Open Source brand also runs Packt's Open Source Royalty Scheme, by which Packt gives a royalty to each open source project about whose software a book is sold.

Writing for Packt

We welcome all inquiries from people who are interested in authoring. Book proposals should be sent to `author@packtpub.com`. If your book idea is still at an early stage and you would like to discuss it first before writing a formal book proposal, then please contact us; one of our commissioning editors will get in touch with you.

We're not just looking for published authors; if you have strong technical skills but no writing experience, our experienced editors can help you develop a writing career, or simply get some additional reward for your expertise.

[PACKT] open source
community experience distilled
PUBLISHING

Raspberry Pi Robotics Projects

Second Edition

ISBN: 978-1-78528-014-6 Paperback: 300 pages

Get the most out of Raspberry Pi to build enthralling robotics projects

1. Make your projects talk and understand speech with the Raspberry Pi.
2. Use a standard webcam to make your projects see and enhance vision capabilities.
3. Full of simple, easy-to-understand instructions to bring your Raspberry Pi online to develop robotics projects.

Mastering BeagleBone Robotics

ISBN: 978-1-78398-890-7 Paperback: 234 pages

Master the power of the BeagleBone Black to maximize your robot-building skills and create awesome projects

1. Create complex robots to explore land, sea, and the skies.
2. Control your robots through a wireless interface, or make them autonomous and self-directed.
3. This is a step-by-step guide to advancing your robotics skills through the power of the BeagleBone.

Please check **www.PacktPub.com** for information on our titles

[PACKT] open source
community experience distilled
PUBLISHING

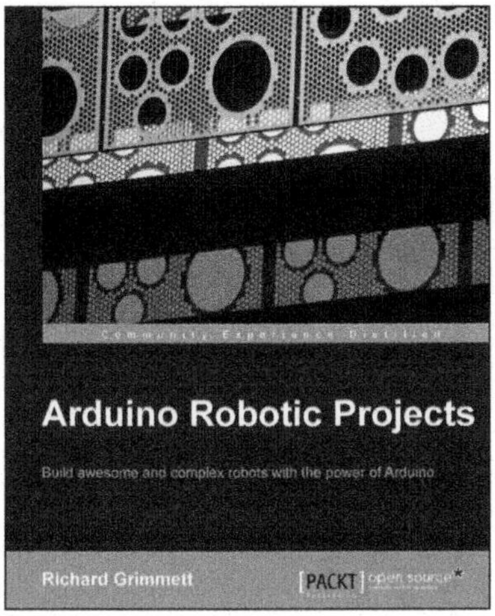

Arduino Robotic Projects

ISBN: 978-1-78398-982-9 Paperback: 240 pages

Build awesome and complex robots with the power of Arduino

1. Develop a series of exciting robots that can sail, go under water, and fly.
2. Simple, easy-to-understand instructions to program Arduino.
3. Effectively control the movements of all types of motors using Arduino.

Building Multicopter Video Drones

ISBN: 978-1-78217-543-8 Paperback: 128 pages

Build and fly multicopter drones to gather breathtaking video footage

1. Capture mind-blowing video from the sky and impress your friends and colleagues.
2. Choose and configure components to build your own multicopter.
3. Fully illustrated and written in a conversational manner.

Please check **www.PacktPub.com** for information on our titles

Lightning Source UK Ltd.
Milton Keynes UK
UKOW05f2150130817
307198UK00001B/29/P